S0-CPE-769

The Heart of the Cornbelt:

An Illustrated History of Corn Farming in McLean County

The Heart of the Cornbelt:

An Illustrated History of Corn Farming in McLean County

William D. Walters, Jr.

McLean County Historical Society
1997

The Heart of the Cornbelt:

An Illustrated History of Corn Farming in McLean County

Library of Congress Cataloging in Publication Data

Main entry under title:

This book was composed by Jane Flanders Osborn of Osborn & DeLong, printed and bound by Pantagraph Printing and Stationery Company, all of Bloomington, Illinois.

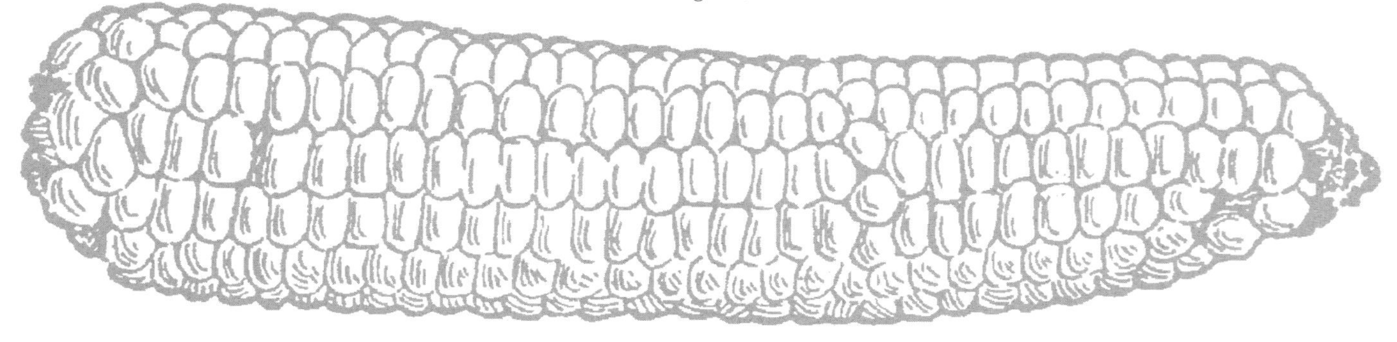

This publication is made possible in part by a grant from the Illinois Humanities Council, the National Endowment for the Humanities, and the Illinois General Assembly.

Table of Contents

The Heart of the Cornbelt:
An Illustrated History of Corn Farming in McLean County

William D. Walters, Jr.

FOREWORD

This book is the story of corn farming in McLean County, which for much of its history has been the leading corn-producing county in the United States. It was developed in close association with the McLean County Historical Society, to be published in association with its "Just Corn" exhibit. The aim of the book is to provide readers with a sense of what corn farming has been like in McLean County. It is not just about corn, but about the people who raised corn, the buildings they lived in and worked around, the problems they faced, the landscapes they created, and how all of these have changed over time. Whenever possible, I have tried to let the farmers tell their own story and I have tried to permit the illustrations speak for themselves; I hope that the readers will regard the pictures as texts in their own right and will spend time thinking about what these pictures have to say.

I have tried to keep references to an absolute minimum. I hope there is enough information provided to start the reader in the right direction, but everyone should be aware that the brief bibliography at the end is just a minute fraction of the vast amount of material available on American corn farming. Many of the sources used are from the archives of the McLean County Historical Society, and if not noted otherwise, this is the source of all manuscript material mentioned. During the spring and summer of 1997, volunteers interviewed dozens of farmers and retired farmers in McLean County. These interviews have all been transcribed and are available in the Society's archives. These interviews provided much of the information which went into chapters VII and VIII. In these chapters, when a farmer is mentioned or quoted, the text of the full interview may be found under the farmer's name at the Society. I have been able to use only a very small part of this rich treasure of information and I hope that everyone will realize how much more is available in the original texts.

Reminders of the past vanish quickly in McLean County. The land is too valuable to permit wasted space, and farmers, even those with a keen sense of their ancestors' contributions, know that farming is first and foremost a business, and a business which can impose merciless demands on the farmland. Twenty-five years ago, I surveyed more than a hundred historically important buildings in the county: barns, corncribs, houses. Today the great majority of these structures are gone. Likewise, the farmers who understood the old ways of farming are rapidly passing from the scene with their stories largely untold. When I speak with farmers, there is often a sense of pride in what they have accomplished, but there is also a sense of regret that no one has been there to tell their story. This book is far too short to be the response these farmers deserve, but it is my hope that one of those who read this book may perhaps be inspired to tell the story in the depth which it deserves. This is also a book about landscapes, and for that I make no apology. I have always wanted to know the meaning and the story of the everyday things which surround me and I hope this book finds others who feel the same way. I hope it is read by people who will then take time to wander down country roads and do their own exploring.

I must thank Greg Koos, who first suggested that I write this book, who did much of the work needed to make it a reality and who has been constant source of knowledge and inspiration. For many years, Greg has been the sounding board for my ideas and my source for many obscure documents. It has been great fun not to agree with him on what is significant. I stand in absolute awe in the face of Greg's great knowledge of McLean County history. Thanks also to Susan Hartzold, who did much of the work in finding and organizing illustrations and who has been largely responsible for the "Just Corn" exhibit. She has, without fail, responded to my requests with intelligence and lively good humor. They, and Pat Hamilton, Librarian/Archivist, at the McLean County Historical Society have found most of the pictures and uncovered many of the manuscripts used in this book. Since I have been permitted absolute freedom in presenting my own interpretations, the views that follow are my own and not those of the McLean County Historical Society. Many, many others contributed to the book. I must thank all of the farmers who were interviewed and all of the volunteers who did the interviewing. Thanks are also due to all of those farmers who patiently permitted me to take pictures and who answered strange-sounding questions with patience, courtesy, and insight. I need also to thank Leo Rennhack and Tim Dargus for sharing knowledge of corn farming. The list of those to be thanked must also include Bill LaBounty for his great work in processing the text and images, Jeff Panfil for finding numerous sources and for making maps, Jo Kimler and Debbie Lescher for frequent secretarial help, the library staff at Illinois State University for finding books. First of America Bank and the Illinois Humanities Council provided generous financial support for the project. There are many others who need to be thanked including Chris Anderson, Kathe Conley, Russell Harris, the Funk Heritage Trust, Rey Jannusch, Ed Jelks, Don Meyer, The McLean County Recorder of Deeds, Ailene Miller, and Duane Sywassink. Of course most thanks of all are due my wife, Karen. She knows that the least important part of her contribution has been the many hours she has spent helping with the manuscript of this book. Without her, none of the good things that have happened in my life would have been possible.

Normal, Illinois, July 31, 1997.

Illustrations

Photograph credit abbreviations: MCHS: McLean County Historical Society; Pant: *Pantagraph*, Bloomington, Illinois; P & B: *Portrait and Biographical Album of McLean County 1887*.

CHAPTER 1

CORN COMES TO McLEAN COUNTY 1200 - 1850

The story of corn in McLean County begins along the banks of Kickapoo Creek. The Kickapoo is not one of the more impressive streams in Illinois, nor does it flow through classic Corn-belt topography. The little river twists between low banks and flows rapidly around gravel bars. The surrounding soil ranges from pale tan in a dry summer afternoon, to Snickers-bar brown in the wet spring morning. This part of the county is not prime corn-growing country. The small tributaries which join the Kickapoo just north of the town of Heyworth are undistinguished. Two of these tributaries, Burlison Creek and Little Kickapoo Creek, might easily be taken for drainage ditches. The land nearby is hilly, and the slopes are now covered with scrubby trees. One could easily mistake a photograph taken here for a representation of one of the less prosperous counties of Indiana. Yet it was just west of Kickapoo Creek, near the point where the Kickapoo is joined by these two small streams, that the only Indian mound in the county once stood. The mound is gone now, plowed down, eroded by rain and dug away by long dead pot hunters; but the pioneer settlers of Randolph Grove knew the mound well and they frequently speculated about its builders.

In June of 1972, a group of field school students from Illinois State University arrived at the mound. At that time, the area was almost entirely agricultural, and the surrounding bluffs had yet to sprout their 1990s crop of wooden boxes for urban refugees. The students, eager to take part in the great adventure of field archaeology, unloaded their tape measures, sextants, trowels and shovels with oddly truncated blades and set out the all-important coolers of soft drinks. In July, they loaded up their gear for the last time, tired, bug-bitten, and wearily aware that archaeology was mostly hard

Figure 1:1. Archaeological field school students at the Noble-Wieting Site 1972. (MCHS)

Figure 1:2. Monument at the Grand Village of the Kickapoo 1911. (MCHS)

work. Among the things the students took away were hundreds of brown sandwich-type bags, each one carefully marked. All contained what were officially called artifacts, but what the director of the field school, Ed Jelks, would often refer to with a smile as "garbage." In some of the bags were bits of fire-blackened vegetable matter, but no one at the time knew they included the remains of the earliest corn ever found in the county. It would be several years before the blackened lumps were sorted, processed by flotation, and identified by experts. The carbonized remains turned out to include remnants of pawpaw, grapes, black walnuts, squash, beans, and what the experts labeled *Zea maize*, Indian corn (Schilt 1977).

The plants found at the mound, officially called the *Noble-Wieting* site, were tiny. Had the ears been complete, they would have measured only an inch to an inch-and-a-half in length. Altogether, there were 140 fragments of corn, representing at least two different varieties. Most of the bits were from eight-rowed ears similar to a miniature version of Northern Flint Corn, which the European settlers knew well and about which a good deal will be said later.

Figure 1:3. Fragments of corn recovered from the Grand Village of the Kickapoo. (MCHS)

A substantial number of the burned fragments came from a more primitive 12-rowed variety of maize. The discovery of the corn was significant because it dated from a time soon after the plant had reached what is now Illinois. Radio carbon tests suggest that the ears were burned sometime around 1200. Perhaps they were remains of a meal scraped into the fire. Most likely, the corn was a garden crop. It was probably planted in small hills, perhaps in the same holes as bean seeds, so as the beans matured, they would wind their way around the growing corn stalk. Squash may have been planted between the corn and bean hills.

All this was, of course, secondary. The people who grew the corn were primarily hunters of white deer and elk, and meat clearly constituted the bulk of their diet. There were, perhaps, fifty people living at the *Noble-Wieting* site who remained for a little less than a century. Their village was a middle-sized place, more than a hunting camp,

Figure 1:4. After leaving McLean County the Kickapoo fled westward. This dwelling was photographed in Brown County, Kansas, about 1906. (MCHS)

Figure 1:5. Mortar for pounding corn used by early settler John Moore. (MCHS)

Figure 1:6 (right). Kickapoo Mkopahmah (Bear Chief) about 1906. (MCHS)

but dwarfed by the vast metropolis which was then thriving at Cahokia. The McLean County people who discarded the bits of corn were the country cousins of the southern Illinois mound-builders. The local natives left no monumental structures, but one likes to think that, at least once in their lives, they followed the streams down to the Sangamon River and from there wandered along the banks of the Illinois to the Mississippi where they could see their kinfolk lugging baskets of dirt up toward the top of Monks Mound.

The only other documented remains of Native American corn growing comes from one of the saddest places in the county. In the years between 1650 and 1813, the Illinois River valley was one of the most dangerous places in the world. It was the scene of repeated warfare, brutal ethnic cleansing, and mass genocide. Many tribes were involved. Iroquois came west to exterminate the Illinois. The Fox joined the French to drive out the Kickapoo. Later, probably in what is now eastern McLean County, the Kickapoo joined the French and others to defeat and slaughter an estimated 900 Fox. Soon after the battle against the Fox, these same Kickapoo moved to the south and settled into what they called their Grand Village, in what is now Old Town Township of McLean County.

The Kickapoo culture was rich in the lore of corn. Once they raised four kinds of the grain, red, blue, white and black. Red corn was *Tacoa a na*, white corn was *Apeskiaki* and black corn was *Makatemina*. Among the Kickapoo of Mexico today, only the black and the red are still grown. The plural of corn was *Medicoaki*, and fresh corn was *Eskicoachiki*. The corn cob was *Owipascoa* and the husk *Okonepacoa* (interview with Miguel Rodriguez, Ekikapoidoani Nocoatoa, 20 May 1997, McLean County Historical Society). Every now and then, some of the modern Kickapoo will return to McLean County to think of those days. This area is still well remembered by the tribe. The years when their ancestors had lived in the Grand Village were perhaps the tribe's happiest years.

In 1790, American forces under Colonel Wilkinson came into the area in an attempt to exterminate the Kickapoo. When they could locate them, Wilkinson's forces burned villages and fields, paying particular attention to demolishing storage facilities and burning supplies of corn. Wherever they went, the Americans left behind terror and starvation. It is uncertain if Wilkinson ever found the Grand Village of the Kickapoo. It may have been missed, surviving for a few more years. Perhaps it was re-occupied after Wilkinson's raid. If so, the Kickapoo endured a few more years of peace before being driven out in 1813.

KICKAPOO WORDS FOR PARTS OF THE CORN PLANT

The Kickapoo people have no written language. The words listed represent the phonetic pronunciations of the Kickapoo spoken word for the parts of a corn plant. These words were provided by Miguel Rodriguez whose Kickapoo name is Ekikapoidoani.

Additional words:

NACOATOA
(Kickapoo)

MAKATEMINA
(black corn)

TACOA A NA
(red corn)

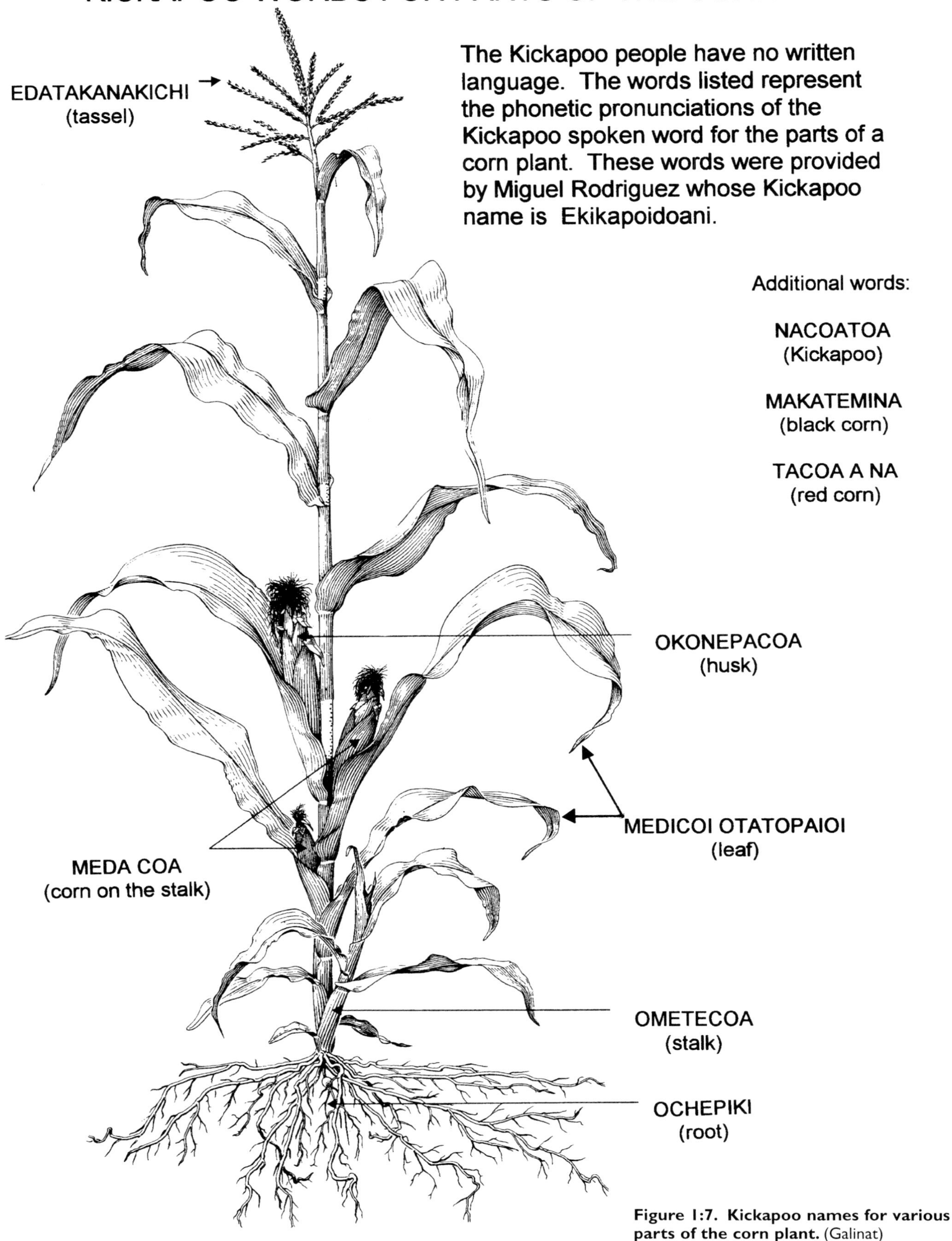

Figure 1:7. Kickapoo names for various parts of the corn plant. (Galinat)

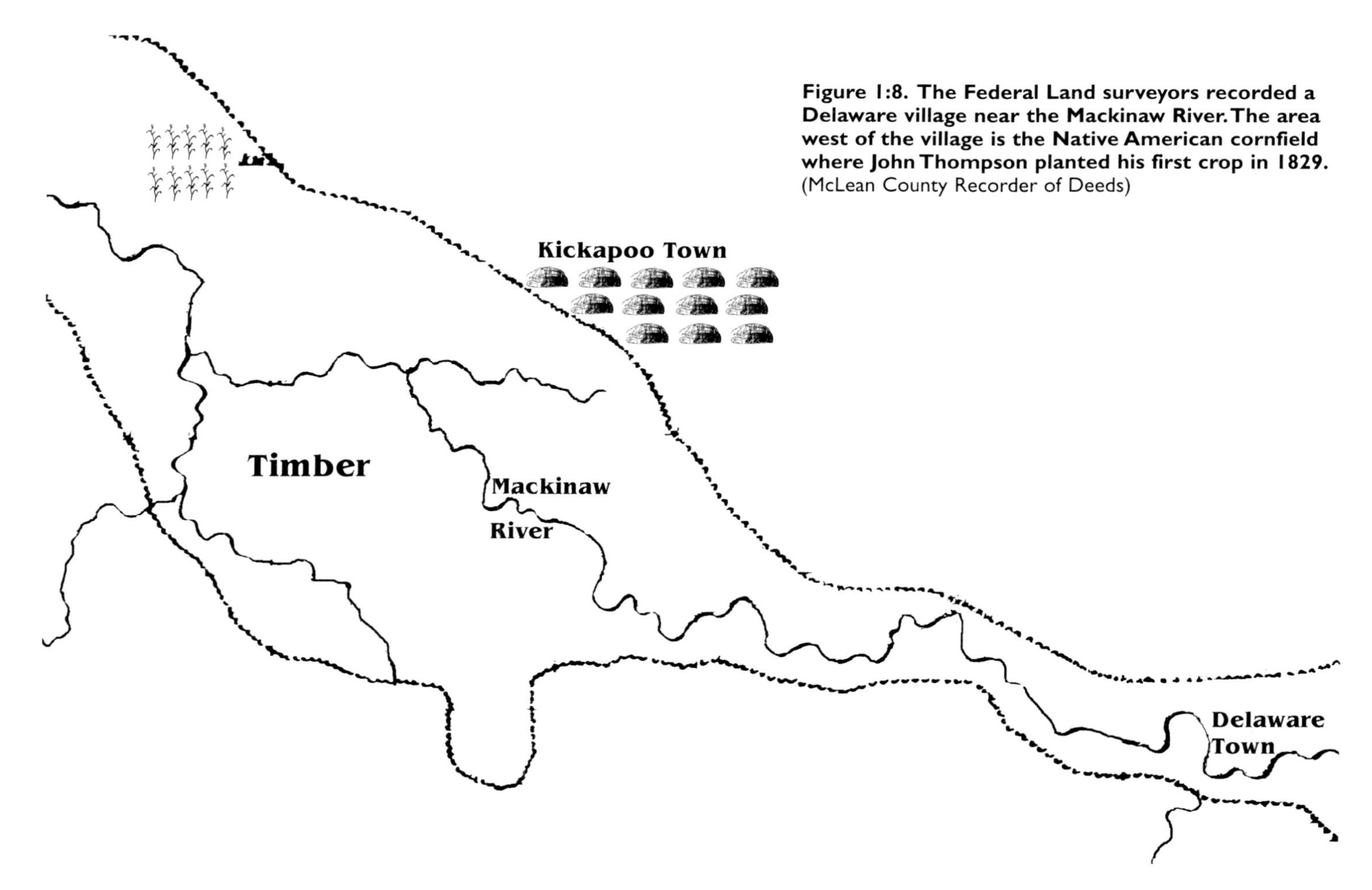

Figure 1:8. The Federal Land surveyors recorded a Delaware village near the Mackinaw River. The area west of the village is the Native American cornfield where John Thompson planted his first crop in 1829. (McLean County Recorder of Deeds)

The war of 1812 brought the end of the McLean County Kickapoo. Troubled by rumors and raids and driven by lust for land, new migrants decided that they could never live in harmony with the Native Americans. Rather than be responsible for the extermination himself, Governor Ninian Edwards called in hired killers from Kentucky, among them future president Zachery Taylor. These Kentucky troops made the location and destruction of the Grand Village of the Kickapoo a prime objective. Initial attempts failed. Finally, in May of 1813, one of Taylor's subordinates located and destroyed the Grand Village. The harried Kickapoo had already fled. Many years later, when the Grand Village was excavated, archaeologists found bells, gunflints, musket balls, kaolin pipes, European-made beads, kettle fragments, scrapers, and remains of a single badly charred ear of corn (Smith 1978).

In McLean County, there were few direct links between Native American corn production and European farming. Settlers got along well with the few remaining Native Americans. Occasionally, however, one finds at least an indirect connection. In October of 1829, 16 years after the burning of the Grand Village, Virginia-born pioneer, John Thompson was scouting for a potential home site. About five miles east of the present town of Lexington, in a wooded area along the banks of the Mackinaw River, he came across the remains of an abandoned Delaware Native American settlement. This was the same Delaware Village that the Federal government surveyors had recorded a few years earlier and which is shown in Figure 1:8. In 1829, many of the lodges were still standing. Thompson used these as temporary stables for his horses and other livestock. He later recalled what he had first found. "It seems that when the Indians were there, they had cut down many trees, for the purposes of burning the tops, and in some places had cut enough to make a little Indian farm or patch for growing corn." Thompson used the abandoned Delaware corn field as a nucleus for his own first crop of corn. He cleared more land and, during the summer of 1830, was able to raise a fair crop (Duis 1874, 137-138).

What procedures did the first European settlers follow in the raising of corn? The first step was to select a site which usually meant choosing a few acres of prairie as close as possible to the margin of one of the groves. The Federal surveyors recorded a few of these fields; they are almost always exactly on the line between timber and prairie. The most favored location was usually on prairie right next to woodland. One early McLean County settler explained that plowing prairie was not easy work, "but it was very easy compared with the labor of clearing timber" (Duis 1874, 618). Studies in several areas of Illinois have confirmed this pattern. Selecting prairie meant that clearing trees was not essential. More importantly, it preserved the much more valuable wooded land for fuel and fencing. The pattern of settlement made maximum use of the available environment. Fence rails were cut in the timber. These rails were used to enclose a small bit of prairie as close as possible to the edge of the woodland. Livestock were turned loose on the

Figure 1:9. The Karr homestead near Bloomington photographed about 1885 given an excellent idea of what pioneer fields and fence were like. (MCHS)

Figure 1:10. Traditional corn-field planting. Note multiple plants per hill and space between hills of corn. (W. Walters)

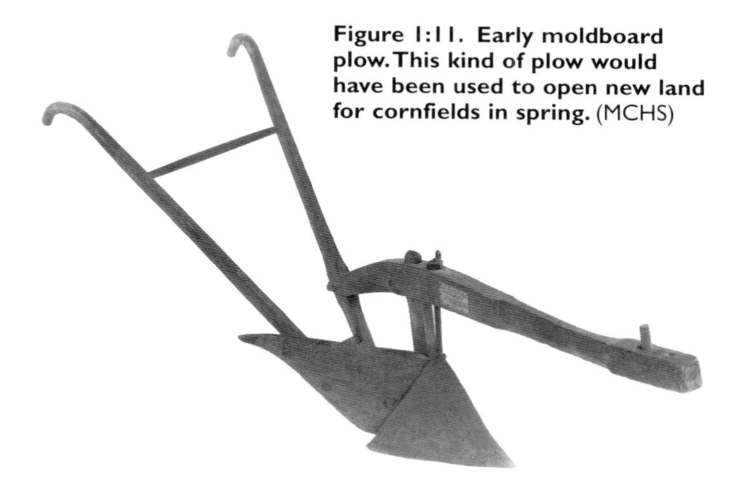

Figure 1:11. Early moldboard plow. This kind of plow would have been used to open new land for cornfields in spring. (MCHS)

Figure 1:12a. Many crib forms were popular. This is a single crib around 1915. (MCHS)

Figure 1:12. Rail cribs were the most common form of pioneer corn storage. This one was still in use when photographed about 1910. (MCHS)

prairie to graze. Everyone kept track of their own animals by a system of ear notching. Before 1840, almost all of the newly established farmsteads were found along margins of groves, but it was not just houses and fields that clung to the edge of the timber. Where possible, roads followed this same boundary between woodland and prairie. All but a few of the newly platted towns also lay astride the line which separated the small islands of trees from the vast ocean of prairie grass.

Many early settlers describe this pattern of land use. When Abraham Enlow grew tired of what he called "grubbing stumps" in Kentucky, he entered 80 acres of prairie and bought five acres of timber to fence his land (Duis 1874, 434). Stephen Taylor entered 80 acres of government land but was forced to purchase 10 acres of timber (Duis 1874, 841). When anyone deviated from this pattern, it aroused comment. Asahel Gridley remembered when Colonel James Latta enclosing 100 acres in what is now Durley's

Figure 1:13. Asahel Gridley, McLean County's first millionaire had much to say on early farming. (MCHS)

addition to Bloomington. Latta's land was located more than a mile from the edge of the timber. "The settlers expressed surprise that Col. should attempt to make a farm so far from the timber" (Duis 1874, 236). Yet Latta's experiment paid off and he was able to raise a successful crop of what the pioneers called sod corn.

These settlers followed the time tested pattern of raising corn. They began in early spring by plowing strips about four feet apart and then cross plowing at ninety degrees to the original furrows. The distance between rows was important because it was dictated by the width of a horse. Cross plowing was essential to keep weeds from choking the corn plants. If the early farmer was fortunate enough to have a harrow, or skilled enough to improvise one, the ground might be further dragged, pulverized, and leveled. Wherever the furrows crossed, small mounds were formed. Into each of these hills, the pioneer corn growers might hoe a little manure, but mostly they counted on the natural fertility of the soil. The hills, each about three to five feet apart, would be the foundation of cornfields for the next hundred years. Even when the earth was not actually mounded, each cluster of seeds continued to be called a hill. Come May, the farmer would drop four to seven kernels into each hill and cover them with a little dirt. He might keep count by reciting the ancient verse, "One for the blackbird, and one for the crow, one for the cutworm, and two left to grow." (Cuthbert 1844, 748). When the young corn plants were about a foot high, the farmer would walk through the field, pulling out the least promising ones so that only four remained in each hill.

Cultivation followed. It was the goal of farmers to control weeds by plowing the fields three times in each direction. One horse plows were the rule. Using such a plow was described as being pretty much like trying to drag a cat by the tail (quoted in Hurt 1982,11). Jonathan Coon, an early settler in what would eventually become Money Creek Township, later recalled the make-shift nature of such a rig. It is interesting to notice how few of the words which this ordinary farmer used to describe his plow are familiar today.

Figure 1:14. Because cattle and hogs roamed free, fence maintenance was an essential task for farmers. Much of the corn farmer's time was spent keeping livestock out of the fields. (*Tim Bunker Papers*)

Figure 1:15. Satire on free ranging prairie hogs. New England sailors have hired a prairie schooner and abandoned their ships and taken up harpooning porkers. (*Prairie Farmer*)

A whole farming vocabulary has passed out of common use. "The plow was the barshear; the horse was attached to it by ropes, which looped over the singletree [the wooden bar which connected the plow shaft to the traces of the horse's harness] and passed from there to the harness to which they were fastened by being tied to auger holes. The hames [the two rigid pieces on the side of the horse's collar] were tied over a collar of corn husks. The backband was leather or coarse tow cloth, and the line was a single rope. (Duis 1874, 616-617).

Sod corn, like that Colonel Latta had planted, was a labor-saving alternative. To produce sod corn, the farmer simply scattered

Figure 1:16. Zigzag or worm fences were common pioneer expedients but consumed much valuable land. Corn could not be raised until the field was fenced. (W. Walters)

Figure 1:17. Detail of plank fence at Karr farmstead. Note the wide range of planks used and the snake fence with riders or braces in the center. (MCHS)

Figure 1:18. Planting Osage orange on the huge Sibley Farms in neighboring Ford County. (*Harpers*)

corn into every third furrow at the time the prairie was broken, allowing simultaneous prairie-breaking and planting (Oliver 1843, 86-87).

Come September, the corn was cut and shocked to allow it to dry. It was then picked and hauled back to the farmstead where the husks would be removed. Because there was no economical way to move corn to market, the corn was either stored in ear form for later use or fed to livestock. In the fall cattle and hogs would be driven to Galena, Chicago, or even Cincinnati. The great period of McLean County livestock drives lasted from about 1830 until the early 1850s, a little longer than the epoch of the much more widely known Texas cattle drives.

An early observer of McLean County farming practices set down his thoughts in the following verses.

Early on the dewy morn,
They turned the soil and they raised the corn;
They gathered the corn and they raised the pork,
They steadily prospered by constant work,
They sold the pork with cash in hand,
Their instant thought was "get more land,"
They got more land and raised more hogs;
They raised more cash, the greedy dogs,
In ceaseless grasp and thankless work,
Purchased the chase — more corn, more pork,
More cash, more land, more corn, and then
More land, more cash for greedy men.
(Quoted in Hardeman 1981, 29).

The combined enemies of early corn growing were hogs and the law. Early hogs were long-nosed, long-legged, lean-flanked savages famed for insatiable appetites and foul tempers. From March to November they wandered across the prairie, great herds terrorizing settlers, devastating crops, and generally making life miserable for crop growers. The difficulties created by roving hogs would have been much less severe had the law not been on their side. Like the mandates of the eastern states on which it was based, and like English common law, Illinois statutes gave priority to free-ranging livestock. Cattle and hogs could roam where they would and eat what they wished, provided that they did not penetrate what the legislators had defined as a "legal" fence. In other words, the law forced the early farmer to surround his crops in order to keep livestock out and permitted the same farmer and his neighbors to use any unfenced land as open range. As might be expected, it was a bitterly disputed law. Agricultural reformers raged against it, crying, "In a land where no fencing material grows, no fence shall be required" (*Union Agriculturist* 1842, 84). But stock raisers throughout the state resisted any changes. They claimed it would bankrupt them to confine cattle and hogs with fences. For thirty years, the battle over fence law

Figure 1:19. Early hedgerows were trimmed at armpit height. This hedge in Old Town Township is one of the few surviving examples of traditional trimming in the county. (W. Walters)

reform raged in Springfield and in the agricultural press, but it was also fought in a more savage way around prairie farmsteads.

The law might require a cornfield to be fenced, but many farmers took matters into their own hands. Lawyers could assert that unfenced prairie was legally common grazing land, but farmsteads were isolated and nights were dark. A man could not stop a hungry sow from ravaging his corn, but he could send a harsh message to the animal's owner. Even after the stock laws went over to a system of township option, the *Pantagraph* recorded many problems. "We are having a little dispute once in a while about stock running at liberty on the commons" (17 June 1874); and four years later; there was an angry dispute last Sunday about turning a calf out on the commons, "which ended as usual in a footrace and no bloodshed" (5 April 1878). The incidents were not always resolved by farmers chasing one another down muddy lanes. Near Clarksville, James Hisel and S. K. Scott got into a spirited fight over the question of hogs in the cornfield. The fight resulted in Hisel "getting a very severe choking." Sometimes the hogs suffered. "Some cruel cuss disemboweled two of G. W. Rees' hogs, either with an ax or a hoe or some other sharp instrument. The hogs were running at large on the commons at the time" (*Pantagraph*, 16 July, 1880).

Given these considerations, it is not surprising that the major topic of the day was how to fence cropland. Fence was the single most important problem confronting early corn farmers. The question of how to fence dominated early Illinois agricultural journals. The problem was simple: a farmer couldn't grow corn, or any other crop, until he had fenced his fields, and there was not enough timber locally available to provide the fencing material. One settler wishing to show off his Latin wrote that the *ne plus ultra* [the summit of achievement] of farming was "a good tract of timber and prairie adjoining" (*Union Agriculturist*, Aug. 1841, 60). Successful farming required a combination of timber and prairie. A rough rule of thumb was that a farmer needed one acre of timber to fence each ten acres of prairie. As a result, prairie was sold in large blocks and groves were minutely divided into woodlots. Even today, one of the most striking features of McLean County land ownership is that former timber is far more minutely subdivided than former prairie. This ownership pattern goes back at least a hundred and fifty years and is directly related to the fence laws.

In the earliest years, the split rail zigzag or worm fence was nearly universal. William Oliver (1843, 239-240) estimated that a farmer could split 100 to 150 rails a day and that a wood worker could erect about 200 yards of such fence in a day. No wonder then that a good rail splitter was always popular. McLean County pioneer Isaiah Coon bragged that he could split 200 rails a day and pocket 50 cents for each hundred, but such a pace was exceptional (Duis 1874, p. 620). Split rail fences continued to be built into the second half of the century and some can still be seen surrounding fields in the 1874 county atlas illustrations. Such fences were picturesque, but they were never popular with Cornbelt farmers. In the first place, everyone agreed that even the worm fence was no match for a hungry hog. Moreover, zigzag fences consumed substantial amounts of land and were hard to work around, especially so as farm machinery became more common. As soon as feasible, plank fences were substituted. Often the farmer would haul wood from his timber to the nearest sawmill and return with a load of boards. In March of 1852, the son of a pioneer farmer entered into a contract with John McClun to rent one of McClun's unfenced farms. Young Orendorff was to get two dollars an acre for fencing the farm. Wood for fenceposts was to come from McClun's timber and Orendorff was to get "what is right and fair" for hauling timber to the mill, hauling lumber back to the farm, and "nailing on the said boards." Ultimately, he received $105.69 for fencing 65 acres, plus $132.43 for breaking the newly fenced prairie. Therefore, on paper, McClun lost money on each acre he rented. In reality, a fenced broken farm was worth so much more than raw prairie that the land owner came out ahead (McClun Papers; Walters and Smith 1992, 17).

Wooden fences, plank or rail, were only practical if the corn farmer had access to nearby timber which limited early farm location to places near groves. What other options were available? Wire fencing was known and discussed in the agricultural press, but practical farmers, like T. Borland of Mt. Hope, found no wire fence available locally (Borland 1852). Much more practical were ditches and hedges. Ditches seemed to many to be a cheap and practical expedient. In addition to being the largest county in the state, McLean is the only county in Illinois with no exposed bedrock. Timber was scarce, but dirt was abundant, easily plowed into banks, and free. One of the many who experimented with such fences was N. E. Hall, who farmed near Hudson. In 1837, he made a ditch three feet deep with an accompanying embankment, "after the fashion of the day," and used it to enclose two sides of a 40 acre field. Hall's embankment was finished off with yellow clay from the bottom of the ditch and supplemented by a rail fence on the outside. The experiment was unsuccessful. The ditch "dried and decayed" and was found to be very liable to be "dug down" by cattle and hogs. Two years later, Hall tried a revised double ditch crowned with a specially designed rail fence, which he found to be very satisfactory (*Prairie Farmer* 1847, 98). Others built ditch and sod fences of their own design. Most farmers, however, found that hedges provided a better answer.

Thorn hedges had their advocates, as did willow hedges for low ground, but the most important hedging material was the

Figure 1:20. This Osage Orange hedgerow in Blue Mound Township has remained untrimmed for many years. (W. Walters)

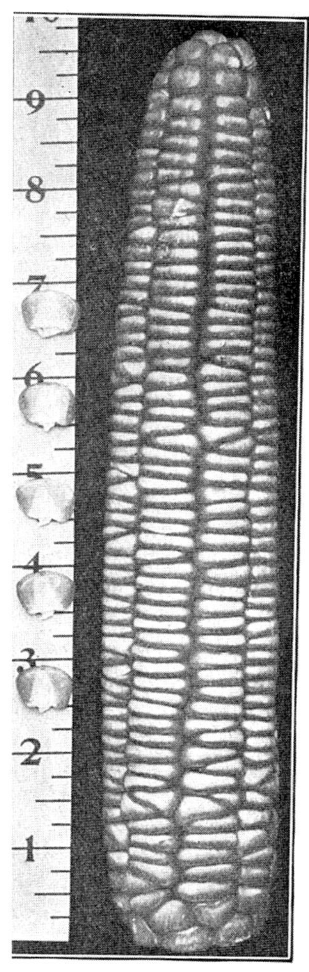

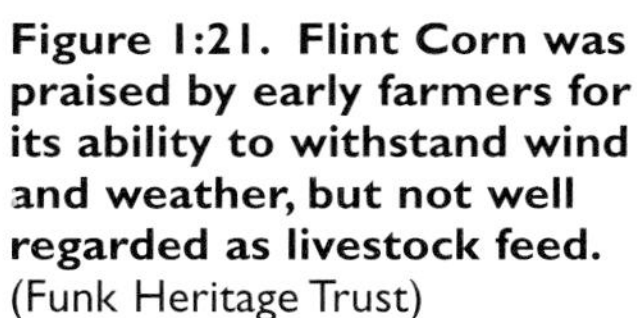

Figure 1:21. Flint Corn was praised by early farmers for its ability to withstand wind and weather, but not well regarded as livestock feed. (Funk Heritage Trust)

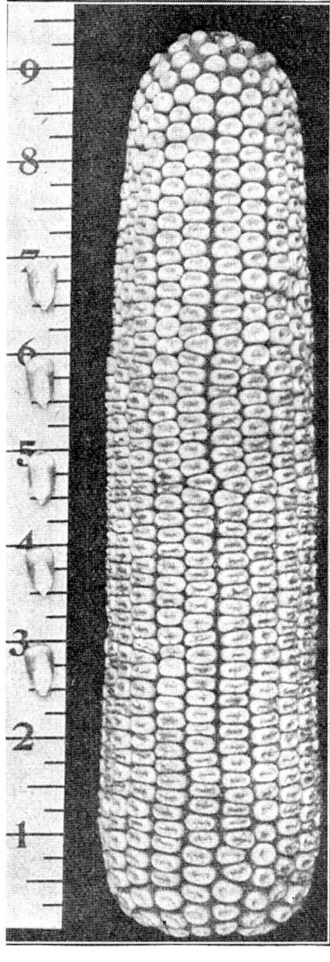

Figure 1:22. Yellow Dent Corn was also well known to McLean County's early settlers. This is Reids Yellow Dent, a later development. (Funk Heritage Trust)

Osage orange. Rail and plank fences have almost completely vanished from McLean County, but mile after mile of Osage orange remains and this is only a small part of the original total. Osage orange is not native to Illinois. Seed was brought up from Texas and, by the late 1840s, a thriving nursery business had grown by selling seedlings to farmers who were tired of the constant maintenance of rail or plank fences. There was also a psychological element to planting hedge rows. Many early farmers came from northwestern Europe, and others read travel accounts or books of agricultural advice written by Europeans. For such people, hedges were associated with civilization. They felt that once fields were hedged, the farmer's problems with fencing would be solved.

Isaac Dement, a farmer living near Stouts Grove, explained how he set about creating an Osage orange hedgerow. From a local nursery, he purchased small Osage orange plants which cost five dollars per thousand, delivered to the farm. These plants were set in a plowed trench six inches apart. Roots were set firmly in the ground, and plants were laid at an angle with their branches interlocking. In the fall of the first year, Dement raised them to a vertical position and covered them with soil for the winter. In the spring following planting, Dement found that each bud had sprouted, multiplying the number of shoots available from the hedge. In summer, when the new shoots were about two feet high, a trench was plowed on either side, and the shoots were trimmed to encourage spreading. He reported, "The gophers worked on them in the fall of last year [1851], and we put some castor beans in their holes, and they soon got sick of their undertaking; the beans grew up among the hedge and did no harm." He filled in the gopher-created gaps in the new hedge by bending the tops of adjacent plants together and staking them down. After the first year, the plants were trimmed in spring or summer. In three years, the young trees were from two to 15 feet high. Dement bragged he had created a hedge that cattle of any description could not be run through (Dement 1852, 81).

What sort of a product were the pioneer farmers growing behind these fences? The simple answer is that it was quite different from modern corn. Were a present day citizen of the county to walk along a country lane of 1840, that person would first be impressed by a lack of uniformity in the corn. Each field would differ from its neighbor in size and color of corn, and within any field there would be a wide range of plant size and configuration. Many farmers had worked hard to improve corn, but they had little success. The best ears of corn, which were selected for planting, did not produce new ears which consistently resembled their parent ears. The visitor would also be struck by the fact that the average plant was somewhat smaller than present varieties. Perhaps the most striking feature would be the much greater distance between plants. The overall effect would be of a field that was irregular, scruffy, and thinly planted.

Certain types of corn were mentioned more than others. Most distinguishable was "Flint Corn." In his widely quoted and generally excellent book, *Eight Months in Illinois*, William Oliver stated that Flint corn was the most widely grown variety in southern Illinois. He thought it was sweeter and more palatable than "yellow corn." On the bottoms, Oliver was told, a man could get 100 to 120 bushels of "white flint" to the acre and that elsewhere, 50 bushels to the acre was reckoned a good crop (Oliver 1843, 85). His informants weren't lying to Oliver, only greatly exaggerating. Most would have been happy with 50 bushels to the acre, but they would never have expected it. Thirty bushels to the acre was much more typical of a successful harvest. Flint corn had eight rows of kernels and a narrow, tapering ear. In Mexico, where it probably originated, it is called *Maiz de Ocho*. Early Illinois settlers regarded white flint corn, with its small, hard kernels, as good for yield, but "slow maturing" and poor livestock feed. One LaSalle County farmer put it this way in an 1852 letter "I don't see how the New Englanders can expect to convert it into pork although ground with economy, unless they can invent a plan to feed one bushel a second time. As for us Buckeyes, Hoosiers and Suckers, we would as quick think of making pork out of hemlock knots and white oak pins as flint corn" (Prairie Farmer, 1852, 129).

Yellow dent corn was radically different. Some authorities

have suggested that Flints and Dents are so unlike that if they were wild plants they would be regarded as different species (Hudson 1994, 50). Yellow dents have short, thick ears with from 14 to 22 rows of kernels, as well as a distinct puckering, or "dent" in each kernel. Like Flint, dent corn was also popular with early farmers. "Yellow corn" yellow dents matured early and seemed to please livestock. Some farmers argued that, because dent corn was more completely masticated, the oil from the corn was more useful in producing meat. Yellow corn also had its detractors; many complained that it provided poor yields. Mixed fields of flint and dent were common. For example Osborne Barnard noted in his diary, for July 4 1860, that he had planted all day, both white and yellow corn. Many other varieties were in use. There was Blue corn, Canada corn, Hoosier corn, Ohio Dent corn, Baden corn, and New Jersey Yellow corn as well as dozens of others. All of this corn is what we would today call "open pollinated," but the term was not in use at the time because farmers could not yet imagine a cornfield where corn pollen was artificially manipulated.

In the 1840s and 1850s, no one was sure what crop would eventually prove most successful in McLean County. Pioneer farmers knew that they had run across some of the best agricultural soil in the world, but they were still uncertain if it would eventually be used to grow corn or wheat. Everyone knew that corn had an advantage as a settler's first crop. It was hardy, tough, and easily stored in makeshift bins. The harvest period provided great flexibility and, if needed, it could be eaten "green" to provide nutrition early in the year. Yet, wheat also did well on the new farms and, to many, it seemed to provide higher returns. Wheat certainly did better in dry years. Perhaps, it was only a question of time and improved transportation before central Illinois would become the center of the world's new wheat producing belt. The question of wheat versus corn would only be answered when McLean County farmers began to migrate away from the margins of the timber and onto the tall grass prairie.

It is surprising how little attention the surviving accounts pay to the corn itself. The earliest settlers were deeply concerned with timber for fuel and shelter. They have a great deal to say about the problem of fencing. They write of cold, hunger, poverty, hunting, and the adventures of travel. They have left behind hundreds of stories about hogs and cattle. There are many accounts of church meetings and political gatherings. There survives a very substantial literature about land purchases and town founding. Yet the crop which sustained them receives surprisingly little mention. It is treated as something that everyone understood and, therefore didn't need to be explained. We know they made corn into hasty pudding, johnny-cake, hot cakes, Indian mush, Indian meal gruel, corn bread, corn cake, corn biscuits, corn crumpets, hominy, green corn dumplings, corn porridge, and summer succotash. We know that, without corn, frontier life would have been much more difficult, but the men and women don't say much about corn itself. The problem is that corn was assumed to be one of the great constants of life. Innovative farm families might experiment with potatoes, try sheep raising, or see how wheat might do on the prairie soil, but everyone raised corn. As a New York agricultural journal put it, "The importance and value of Indian corn are too well known to every agriculturist in this country to need illustration" (Dement 1853, 325). It is also true that when many pioneer memoirs were set down in the 1860s and 1870s, corn growing had not changed much from the way it was done in the earliest years. No one wanted to hear stories about what was still an everyday experience.

In the two decades before 1850, great changes were made in McLean County. Its boundaries were fixed, a dozen towns had been established, and the population exceeded 10,000. Agriculture had taken root, but it was an isolated, pioneer sort of farming. Much remained to be done. Almost all of the of the settlements were within a mile or two of the margins of groves. Most of the land was still unbroken prairie grass. Much of the best potential agricultural country was too wet to farm. Transportation was primitive. There were no railroads. Many strips of land had been legally designated as roads, but almost nothing had been done to improve these strips of land. As one traveler put it, there were no roads, only places for roads. Most of the year, cattle and hogs roamed free on the open land and, because of this their quality could not be improved. There was no economical way to move bulk corn or any other grain to market. There were steam driven grist mills and some water powered ones, but they ground mainly for local consumption. Since shelled corn did not store well, it was processed only when it was needed on the farm. The challenge of fencing had been met by various improvisations and experiments. Hundreds of miles of hedgerow had been started, but it would take years to mature and everyone knew that better solutions to the fence problem were needed. The high cost of timber and the general lack of cash kept farm buildings small. As yet, no one had discovered a way to make meaningful and consistent improvements in corn, and it remained essentially the same plant that the Pilgrims had encountered. In 1850, an honest observer of the county would have had to report that McLean County was still frontier country. Like everywhere else on the American frontier, corn was important; no one had yet coined the term Cornbelt, and certainly no one imagined that events in the obscure wilderness of Illinois would ever be of importance to the rest of the world. Only in the next decade would any of this begin to change.

CHAPTER II

CORN MOVES ONTO THE PRAIRIE
1850 - 1860

The first great step in the transformation of McLean County into the heart of the Cornbelt took place in the decade of the 1850s. No other ten-year period in the history of the county comes close to rivaling this decade as a period of fundamental change. In 1850, the population of McLean County was just over 10,000; by 1860 it was 28,772. The population of the largest town and county seat, Bloomington, increased five-fold during these same years.

In 1850, much of McLean County's corn was hauled by wagon to Pekin to be shipped down the Illinois and Mississippi Rivers to New Orleans, which was then the port for much of the American Midwest. The only real alternative was to feed the corn to cattle or hogs that were driven overland to market. In 1850, Chicago shipped only 362,013 bushels of corn. In 1860, it shipped 13,700,133 bushels, and this total included a large part of the grain shipped from McLean County. Changes in this decade were radical and fundamental. This chapter will explore several closely related themes: the continued agricultural development of McLean County, the movement of farmers from the groves out onto the prairie, and the roles played by the different ethnic groups which flooded out into the county.

It must first be remembered that McLean County, in 1850, was still largely frontier country. It was a place of readily available land, homemade farm tools, and isolation from the nation's markets. For many people, the corn farming techniques of the 1850s were not much different from those of the 1830s and 1840s. The term pioneer might be defined as a person for whom improvisation was still a critical skill for survival or as a person whose existence revolved around the need to deal with unfamiliar environments. By either of these definitions, McLean County in the decade of the1850s was still a place of pioneers. The new environment they struggled with was not defined in terms of distance from the east coast, but in terms of vegetation. In many ways, the decision to leave the sheltering neighborhoods of the groves was as tough as the decision to leave Vermont

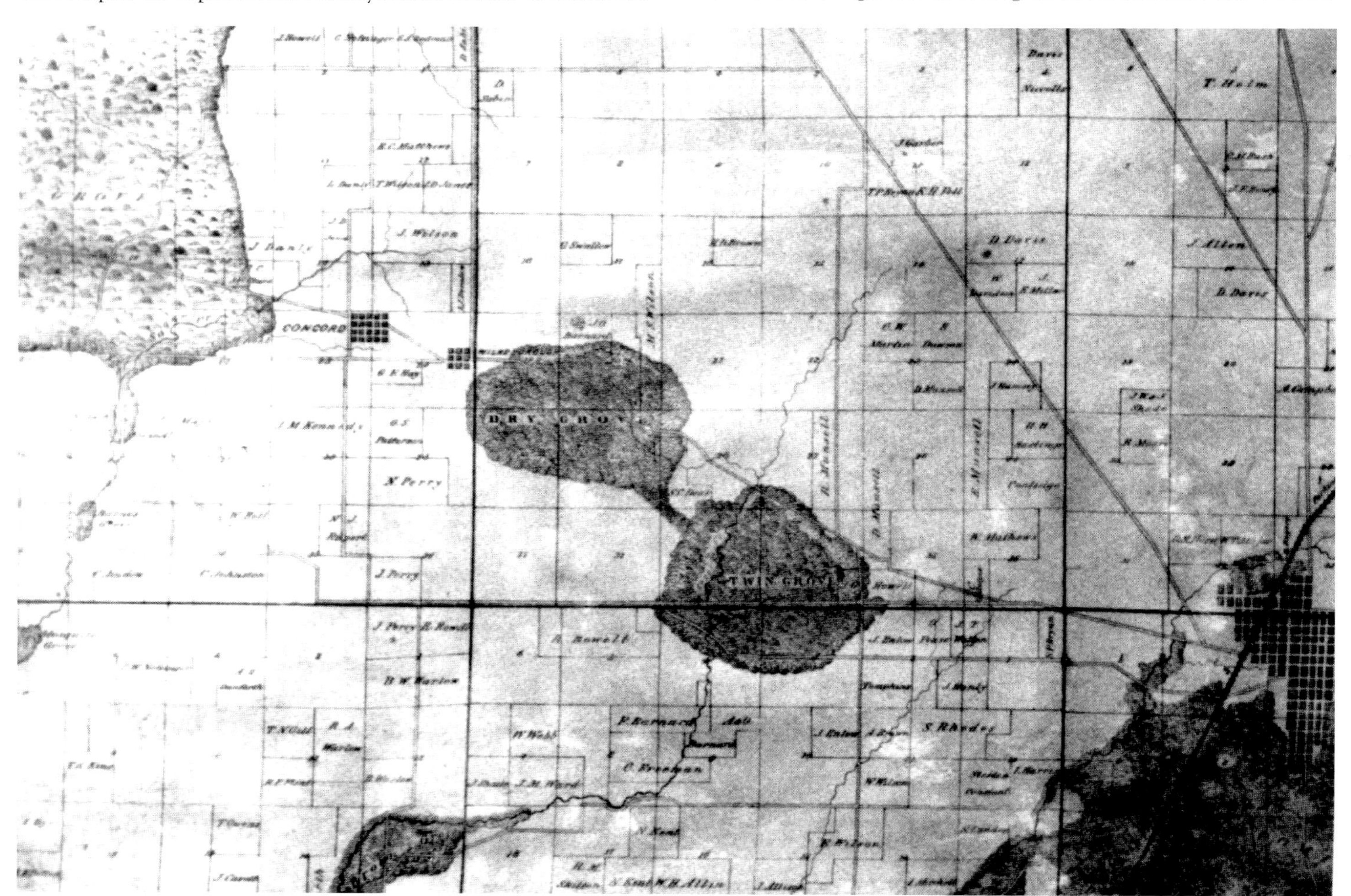

Figure 2:1. Before settlement eleven percent of McLean county was timber and the remainder tall grass prairie. (MCHS)

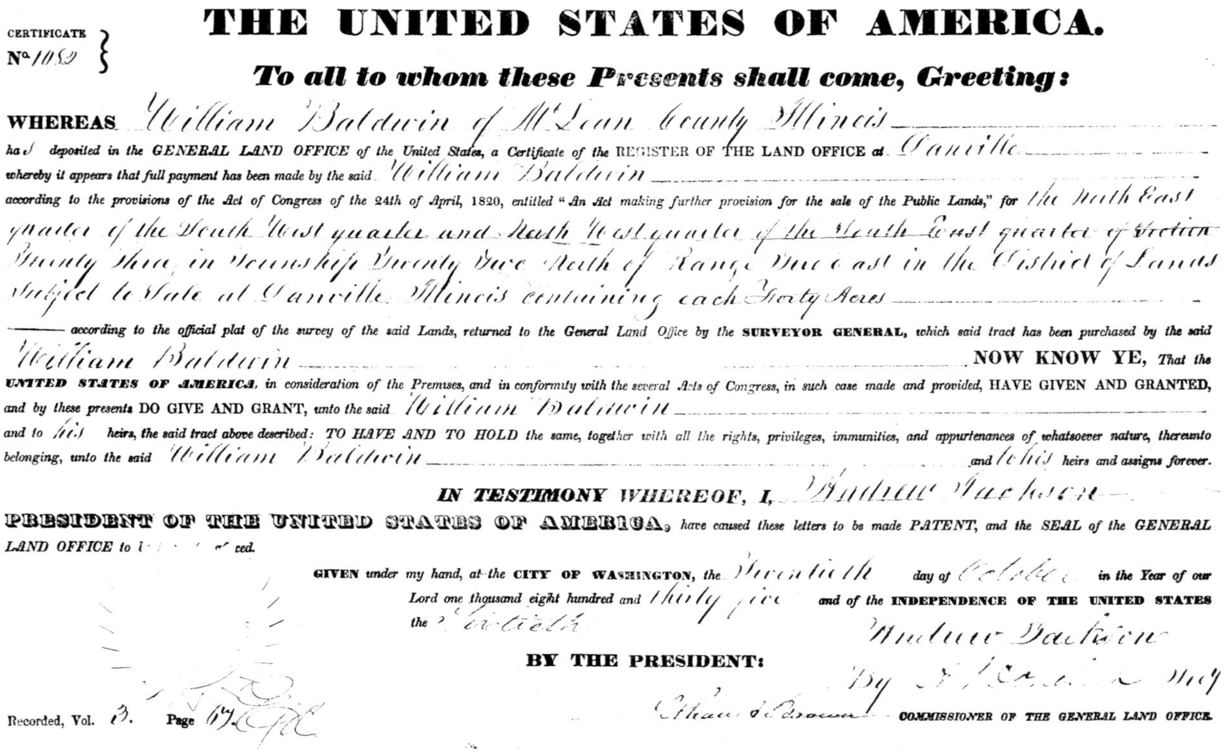

CERTIFICATE
No. 1082

THE UNITED STATES OF AMERICA.

To all to whom these Presents shall come, Greeting:

WHEREAS William Baldwin of McLean County Illinois has deposited in the GENERAL LAND OFFICE of the United States, a Certificate of the REGISTER OF THE LAND OFFICE at Danville whereby it appears that full payment has been made by the said William Baldwin according to the provisions of the Act of Congress of the 24th of April, 1820, entitled "An Act making further provision for the sale of the Public Lands," for the North East quarter of the South West quarter and North West quarter of the South East quarter of Section Twenty Three, in Township Twenty Five North of Range One East in the District of Lands Subject to Sale at Danville Illinois containing each Forty Acres according to the official plat of the survey of the said Lands, returned to the General Land Office by the SURVEYOR GENERAL, which said tract has been purchased by the said William Baldwin NOW KNOW YE, That the UNITED STATES OF AMERICA, in consideration of the Premises, and in conformity with the several Acts of Congress, in such case made and provided, HAVE GIVEN AND GRANTED, and by these presents DO GIVE AND GRANT, unto the said William Baldwin and to his heirs, the said tract above described: TO HAVE AND TO HOLD the same, together with all the rights, privileges, immunities, and appurtenances of whatsoever nature, thereunto belonging, unto the said William Baldwin and to his heirs and assigns forever.

IN TESTIMONY WHEREOF, I, Andrew Jackson PRESIDENT OF THE UNITED STATES OF AMERICA, have caused these letters to be made PATENT, and the SEAL of the GENERAL LAND OFFICE to be hereunto affixed.

GIVEN under my hand, at the CITY OF WASHINGTON, the Twentieth day of October in the Year of our Lord one thousand eight hundred and thirty five and of the INDEPENDENCE OF THE UNITED STATES the Sixtieth

Andrew Jackson

BY THE PRESIDENT: By [illegible] Secy

Ethan A. Brown COMMISSIONER OF THE GENERAL LAND OFFICE.

Recorded, Vol. 3. Page 672

Figure 2:2. Richard Baldridge used land warrants similar to this one to purchase his first McLean County farm. (MCHS)

or Virginia and head west. One can follow this move onto the prairie by looking first at the farming career of Richard Baldridge.

Baldridge was a farmer, school teacher, and self-taught surveyor. He grew up in Adams County, Ohio, working on his father's farm. At the age of twenty-one, when, by law and tradition, a young man no longer owed his father labor, Baldridge married and set out for the west, taking with him little more than a new wife and a single horse. The couple reached McLean County in April of 1852. At this time, well before the Homestead Act, the purchase of even a modest farm was well beyond the means of men like Baldridge. He did what many a young man with more ambition than capital has done before and since; he looked for a place where skill and hard work could be substituted for ready cash. Baldridge was fortunate to find such a place in what is now White Oak Township. This was a rugged, wooded area and because of this had been settled early. Baldridge looked for and located a farm in need of a tenant. The land he found had already been plowed and marked off, but there was no crop in the ground. He quickly agreed to terms with the land owner. Baldridge was to be furnished with seed; given the use of a 16-foot-square room, the best in the house; and loaned a second horse when a team was required for double plowing. He was also to have the use of a two acre garden. Rent was to be "half the crop thrown in the crib." Nearly 150 years later, half the crop is still the common rent on many McLean County farms. In 1850, some farmers were already using mechanical corn planters. Baldridge had nothing of the sort, but "my wife knew how to drop the seed in the ground as I marked it off with a single plow, and the second time round I covered the corn with a shovel plow. Got a good crop" (Baldridge Memoirs).

In the next year, 1853, Baldridge set out to look for a new but similar situation. This time, he learned of a nearby widow who had a room to rent and land which she could not cultivate. Richard Baldridge took it and again received half the crop for his efforts. In 1854, he had saved enough to purchase 80 acres at four dollars an acre in land warrants. About 15 acres of the new farm was what Baldridge called "hazel land." By this, he meant that it was scrub on the margins of the timber. The remainder was prairie and agriculturally much more valuable. He put 55 acres under cultivation and proceeded to acquire the needed buildings. Baldridge constructed a frame house and traded for a log stable which he moved near the house. Unfortunately, a sudden prairie fire soon burned down the stable. Another neighbor gave Baldridge, in return for grubbing up trees, an old barn which was also moved onto the farm. He was able to earn a little more money by teaching school in the winter. Baldridge had moved a step up from renting, but he was still farming in a traditional kind of location: land within a few hundred yards of the edge of timber.

Richard Baldridge's time of decision came in the fall of 1855. The young farmer found that he was able to sell his existing farm and purchase one more than twice its size in Hudson Township. The new farm was almost entirely grassland, and Baldridge soon came to call it his "Prairie farm." Prairie farming was entirely different than working land at the edge of timber. Because the farm was without woodland, he was forced to also buy a separate ten acre

Figure 2:3. Oxen like these on a turn-of-the-century McLean County farm were slower than horses but more suited to heavy tasks like prairie breaking. (MCHS)

wood lot in the nearest grove. He then confronted the great challenge of the 1850's, breaking prairie. In partnership with his father, he purchased a yoke of young cattle and, as he puts it delicately, "made oxen of them." His brother-in-law added three yoke of big steers and the family had soon broken 86 acres of land, which they planted in spring wheat. Disaster followed. The wheat froze and Baldridge recalled, "I didn't get a handful." He did well with a replanting of fall sown wheat, but decided that in the next year, he would put "most of my broke land" into corn. It was another disaster. The year was dry and "I got next to no corn, only 6 or seven

Figure 2:4. Ox shoes like those worn by Baldridge's Buck and Berry differ considerably from those fitted to horses. (MCHS)

bushel per acre." Still, Baldridge persisted. In the spring of 1857, he broke 45 more acres and the following year, still more. By this time, his oxen were well used to their task. Buck and Berry were his favorites. Baldridge would call, "Ho Buck," and the first ox would lumber out of the rail corral and stick his neck out to be yoked; then Baldridge would shout, "Come under, Berry," and the second ox would join its companion. Baldridge, who had always been a staunch anti-slavery man and who had already had a number of narrow escapes while conducting slaves northward on the underground railroad, left to fight in the Civil War. By then, his confidence in prairie soils was complete. In 1864, while still on active service, he instructed his father and brother to purchase 160 acres of raw prairie near the new railroad town of Chenoa for $12.50 an acre. Before he returned, Baldridge's brother had broken 55 acres on the new farm, and it was soon in profitable operation. In a little over twelve years, he had advanced from a renter on the brown soil of White Oak Grove to the owner of two prime farms cut from untouched prairie.

Many myths have grown up about prairie breaking. Therefore, it is perhaps best to try to understand it by looking at the account of a farmer who was actually involved in the process. One of Baldridge's neighbors, N. F. Hall, lived near Hudson, and he has left a detailed account of his prairie breaking experiences. His remarks were set down in the fall of 1847, but they reflect back on fifteen years of prairie farming. Hall was convinced that for breaking prairie, the "new" plows were no better than the old wood and iron devices. He felt that a good prairie breaking plow for use with oxen should

Figure 2:5. Corn cribs like this were built for over a hundred years in McLean County and were among the most valuable structures on many farms. (W. Walters)

be built around a nine-foot-long central beam which should be "as large as a barn sill — five inches square from the bolt to the cross beam — as the beam will twist like a rope on occasion of plowing hazel, rough or large roots, unless very strong." Hall strongly recommended that the ideal prairie breaking plow have both trunk wheels and a coulter, that is, a vertical iron blade which traveled in the ground. As he expressed it, "the coulter is the rudder of the plow." The cross beam, he thought, should be three feet long, four inches thick and ten inches broad. The wheels ought to be three and a half inches thick and 15 inches in diameter. The right wheel ought to run in the furrow with an iron plate fastened to the beam, lapping over it to act as a scraper. This wheel should take most of the plow's weight. Hall strongly recommended iron bands and boxes for the wheels.

To pull his massive prairie breaking plow, Hall used six yoke of "rather light" oxen. With the help of a fifteen-year-old boy, he was able to break 200 acres "almost entirely done from May 15 to July 15, mostly done at from 4½ to 5½ acres per day, consisting of

Figure 2:6. Bins at either side of corn cribs were designed to hold ear corn. Slatted sides permitted circulation of air which helped to dry the corn. (W. Walters)

Figure 2:7. Many kinds of prairie breaking plows were in use in McLean County. (*Prairie Farmer*)

small jobs in different places and satisfactory to employees." The same oxen, Hall reported, "done as good as break 100 acres before May 15." When the summer breaking was done, he claimed they were in good enough condition to break another 100 acres that winter. Not everyone agreed that the 12 oxen Hall suggested were required. In reply to a farmer's question, the editors of the *Prairie Farmer* wrote, in May of 1854, that little prairie was broken by plows drawn by more than two oxen (14:5, p.192).

As farms moved away from the margins of groves, prairie breaking became increasingly important. Nothing could be more incorrect than to regard the conversion of McLean County prairie into farmland as the result of a single episode. Substantial amounts of prairie were broken in the 1830s, but the process continued in various parts of the county for another fifty years. At the time of the Civil War, prairie hay was still being harvested, and farmers would still speak of going "out on the prairie" to hunt. When Asahel Gridley, Bloomington's first millionaire, died in 1880, his property still included some unbroken McLean County prairie.

As they moved further out onto the prairie, farmers encountered more and more problems with standing water. The wetness of the tall grass prairie had long been evident. When Federal government surveyors both measured the land and made notes on its quality, two facts quickly emerged: the surveyors knew the land was rich and they knew it had serious problems with drainage. On February 5, 1824, E. Rector set down the following comments about the north central part of what would eventually become McLean County. "This land flat with considerable water now standing." and, "This mile gently rolling, rich soil, some small ponds." and, "This mile gently rolling, rich prairie." Other days' comments are similar. "This mile level rich prairie, some places inclined to be wet." "Rolling rich prairie." "Land good prairie fit for cult.[ivation]." Fit for cultivation yes, but only if one were to risk seasonal problems with standing water. Before the 1870s, there was no economically practical way to remove standing water in a wet year. Some experiments in drainage were tried, without much success. What the farmers did learn was that on wet prairie land, corn was a far better crop than wheat. In the 1850s and 1860s, as more and more farmers moved out onto the prairie, more corn and less wheat was grown. In a very real sense, damp soil was midwife to the birth of the Cornbelt.

Slowly, the rest of the United States was beginning to realize that central Illinois was more than just another backwoods frontier. This realization was fanned by enthusiastic letters written by new arrivals. On January 6, 1856, Horace S. Warren wrote to a friend, "What shall I say of the great West. The land surpasses anything that I ever saw." The loam on the prairies, he reported, averaged three to four feet thick. Warren exulted that the soil was "deep black and rich as any garden. Farms from 100 to 1,000 acres [with]

Figure 2:8. Prairie breaking went on for nearly 50 years in McLean County. Many different systems were used and multiple yoke of oxen were common. (*Prairie Farmer*)

not one foot of waste land and will yield from 25 to 35 bushels of corn to the acre without manure." As soon as he could, Warren told his friend that he was going to go into farming operations 16 miles south of Bloomington, one mile from the new depot on the St. Louis railroad (MCHS archives).

Warren's letter was personal, but there was also a flood of intentionally promotional literature. In 1859, James Caird, a British member of Parliament, was persuaded by Bloomington banker and land speculator, Asahel Gridley, to visit the county and to publish his observations. The aim of the resulting pamphlet was to persuade English investors to put their money into McLean County land. Caird recorded how he was met at the station by Gridley and how the men had ridden out onto the prairie. That year, the wheat crop had been a total failure. One farmer told Caird that on his 2,500 acre farm, he had averaged 30 bushels of wheat to the acre but received only six shillings per bushel. Wheat was not the crop for McLean County; Indian corn was the thing to plant. Caird interviewed another farmer whose reliability was supported by the fact that he was a graduate of Yale. Six hundred and fifty acres of land would cost £3 (about $12.50) per acre. Prairie breaking should be contracted out, which would cost £260. Seed for wheat would be £260. And fence, £240. Cultivating, threshing, harvesting and delivery would cost £500. Thus, at 20 bushels per acre, a farmer, by spending £2,440 could expect a clear profit, which in dollars, would amount to more than $5000. Caird never explained why,

8
O.L.
East on a random line along the South
side of Sec 35. T25 N. R 1. W.
40 00 Set a tempy. 1/4 post.
80 00 Set a tempy sec. post.
Land this mile rich rolling prairie.
West corrected the line along the S. side
of Sec 35 T25 N. R. 1 W.
40 00 Set a 1/4 post in m—
80 00 Set a sec post in m— cor for Secs
34 & 35 T. 25 N. R. 1 W.

East on a random line along the South
side of Sec 36 T25 N. R. 1 W.
40 00 Set a tempy 1/4 post
80 00 Intersected the Merd line S. of the cor
60 lks.
Land this mile same as last.

Figure 2:9. Federal surveyor's notes provide the earliest detailed record of McLean County land. Most comments on prairie are favorable. The sections mentioned here were later detached from McLean and placed in newly formed Woodford County. (McLean County Recorder of Deeds)

Figure 2:10. The restored Patton cabin in Lexington. As soon as possible settlers replaced these with frame structures. (MCHS)

after recommending corn over wheat, he used wheat to calculate profits. One hopes that Card's readers didn't invest all of their savings in anticipation of such returns, but the ratios are probably about right. There was money to be made from the McLean County prairies.

Potential for profit attracted people from many areas. McLean County lies at a cultural crossroads; between 1830 and 1870, it was a place of mixing among New England, Middle Atlantic and Upland South cultures. Did farmers drawn from different parts of the eastern United States farm in significantly different ways when they reached the emerging Cornbelt? In the 1980s, there was a flurry of academic and popular scholarship which stressed ethnic differences. This emphasis on diversity spread to the literature on farming and culminated in Brian G. Cannon's plea to understand the role of American subcultures on the agricultural frontier (Cannon 1991). No one doubts that early in American history, distinct national groups had distinctive ways of farming. Colonial Germans, Scots Irish, and Moravians all had significantly different agricultural preferences and methods. Certainly, the early French colonies in Illinois and Missouri practiced a common field-based agriculture which was radically different from other communities. Did American groups from different areas follow different corn raising practices?

Where did the 1850 farmers of McLean County come from? Greg Koos has recently completed a study of ethnic origin and farming practice in Illinois (Koos 1995). He finds that McLean County farmers were overwhelmingly native-born Americans. Kentucky as a place of origin led with 23 percent, followed by Ohio with 20 percent, and Virginia with 13 percent. However, these figures can be very deceptive if we accept stereotypes. We must not picture Virginians as plantation-owning cotton growers nor should we imagine Kentucky settlers to have been mountain people. It was the lush lowlands of central Kentucky, the Blue Grass and the northern Nashville Basin which supplied the largest number of McLean County immigrants. The Virginians were drawn from the Shenandoah Valley, and the natives of Ohio came mainly from the lower Miami Valley and Pickaway Plains. All of these are rich farming areas with income and agricultural production well above the national average. The Southerners were from the Upland South in the sense that they were not from the low country of the Atlantic Coast, but not in the sense that they were from mountainous areas. Moreover, intensive corn and livestock raising was already being practiced in all of these areas. This is not to suggest that the individual migrants were rich; many brought next to nothing. However, it does suggest they had witnessed relatives and neighbors do well with new methods of corn-centered farming that would have seemed quite odd to their colonial ancestors. In addition, other eastern states were also provided significant numbers of settlers.

Figure 2:11. The rear of the Isaac Funk Home built about 1842. This is the kitchen and service area of a typical Illinois I-House. (MCHS)

John Hudson has attempted to trace the origin of the American Cornbelt. He argues that its most distinctive characteristics were first evident in the Virginia Military District of Ohio, that is, the area in the south central part of the state. "If one has to identify a place where the Corn Belt begins, no better choice could be made than the broad plain of the Scioto at its junction of Paint Creek at Chillicothe" (Hudson 1864, 66). Central to the Ohio system was the feeding of cattle outside on unhusked corn with hogs to clean up the offal. From early on, it was an export-designed, market-based farming system. It is certainly not coincidental that so many early McLean County farmers came from this area.

What about those from other areas? Significant numbers migrated from New England. Did these people really favor wheat because they liked wheat bread while the families from Kentucky raised corn because they ate cornbread? The best way to answer this question is to look at the 1850 census of agriculture. It clearly shows that, regardless of native state, most McLean County farmers farmed in essentially the same way. They all raised about the same mix of crops, fed the same kinds of livestock, and obtained about the same yields. One possible exception being the small number of New York and New England farmers. These Yankees, those the *Chicago Weekly American* called "a shrewd, selfish, enterprising, cow-milking set of men" (quoted in Hudson 1964, 62), did produce slightly more than the average amount of wheat (Koos, 34-38). Still, it should be stressed that even Yankee migrants to McLean County relied overwhelmingly on corn. In short, regardless of where the farmers came from, they very quickly adopted a common system of farming which was suited to local soil and climate.

The 1850 census of agriculture may also be used to form a picture of a typical McLean County farmer. The farmer was forty-one years old, had resided in Illinois for seven years, and had in his household five other individuals, including farm hands. He was worth about one thousand dollars, and almost all of this wealth was in the form of land and livestock. The family farm contained, on the average, 160 acres of which 80 acres was "improved." Improved, in this case, means broken and fenced. Most of the remainder would have been unbroken prairie. By 1860, the proportion of improved land had risen, but there was still much open range. The family farmed

Figure 2:12. The front of the Isaac Funk Home. This is the formal part of the I-House. Note the symmetrical placement of windows and two front doors. (MCHS)

with four horses and had charge of four cows. Very few had mules or oxen, but farms averaged 26 head of hogs. There were a few large flocks of sheep, but harsh winters and roving bands of semi-wild dogs made extensive wool production difficult. There was pasture and there were gardens for vegetables, especially potatoes. Still, the really telling figures are those concerning the production of grain crops. In 1850, the typical McLean County farmer raised 40 bushels of wheat, 100 bushels of oats, and 1000 bushels of corn (Koos 1995, 33).

This picture can be amplified by what we know from the study of material culture. A good deal is now known about the typical corn farmer's McLean County house. Log cabins were built in McLean County, but they were seen as temporary structures and were rarely occupied by successful farmers for more than a few years. Almost none were built after the mid 1850s, when railroads supplied plentiful pine lumber which could be used to make better houses at very reasonable prices. Many cabins survived as sheds, makeshift corn cribs, or dwellings for hired hands. One such cabin, long since relegated to such second line duties, may be seen in the background of the 1874 illustration of the Dennis Kenyon home.

Much more common are structures which have come to be called I-Houses. The term comes from the cultural geographer Fred Kniffen, who instantly recognized them as intrusive elements when he was doing field work in Louisiana; he named them I-Houses because they were like those he knew in Indiana, Illinois, and Iowa. The photographs of Isaac Funk's house give an excellent idea of what such a dwelling was like. Isaac Funk came to McLean County from Kentucky in 1824. The house shown was built about twenty years later. Look first at the rear view which is the business end of the house and contained the kitchen below and perhaps a loft above. This loft was the sleeping place for servants, visitors of lower status, or perhaps teenage males — anyone who was not quite fit for polite society. Usually, the loft had its own staircase leading down to the kitchen, and there was no direct connection between it and the upper floor of the front part of the house. The porch off the kitchen was the center of much of the farm family's domestic work. Except on special occasions, casual visitors entered the house through the kitchen.

Now, examine the front view of the Funk house. This is the formal part of the house, the part most like drawings one would find in contemporary pattern books. It is symmetrical and placed so it can be easily seen from the road. In effect, it is an advertisement of the farmer's taste. The rooms in this section would be a parlor and dining room below and bed chambers above. Notice that the Funk house has one front door for each room. American farmers were slow to adopt the central hallway which fashionable design dictated; they seem to have regarded it as a waste of space and an impediment to ventilation. Equipping the house with two front doors may also reflect the Funk's time in Kentucky, where paired doors are common. By 1850, central hall and single doors had become the rule in McLean County I-Houses.

The I-House was a frame structure with a structure much like nineteenth century barns. It was also a marvelously functional sort of building, especially so for farmers. Both the front and rear wings are only a single room wide, permitting a maximum of natural light and ventilation. Ventilation was critical because cooking and heating were primarily done by stoves and, although they didn't know the reason, nineteenth century farmers knew that poorly ventilated rooms caused illness. Farming is a muddy business. Two part design of the house permitted the farm wife to keep nicer furniture and decorations separate from some of the inevitable dirt and grime.

Hundreds of I-Houses were built in McLean County. They became the standard dwelling of those upper and middle class farmers from Virginia, Kentucky, Tennessee, and much of Ohio. Farmers with less money built single story houses with the same basic design.

New Englanders who settled in McLean County built a slightly different kind of farmhouse. Look at the pictures of Dennis Kenyon's home in Mt. Hope Township. Kenyon's father came to McLean County from Kent County, Rhode Island, in 1840 (*History of McLean County* 1879, 293,581-582; *Pantagraph* 18 October 1895, 7; 7 December 1898, 7).

Like many farmers from New England, New York, or northern Ohio, Kenyon brought with him ideas about how a house should be designed. Like Isaac Funk, Kenyon built a house framed in heavy timber and, like Funk, he kept the kitchen wing lower in height and distinct from the formal part of the house. The difference is that rather than place the kitchen wing at the rear of the house, Kenyon placed it on one side so, like the main part of the house, the kitchen also faced the road. There are many variations of this New England design which have now been given colorful names like Upright-and-Wing, or Hen-and-Chicks. Travel today through unspoiled parts of the rural Midwest, which were settled by New Englanders — Southern Michigan, Wisconsin, the northernmost counties of Illinois; and many houses of this sort may be seen. These differences in house design were fairly short-lived. By 1875, it

Figure 2:13. Isaac Funk was the patriarch of the large and influential Funk family which played such an important role in McLean County agriculture. (MCHS)

was becoming difficult in McLean County to tell farmers' states of origin by looking at the way they built their houses. One suspects that it was also becoming difficult to tell New Englanders from Upland Southerners by dress, accent, or the way they farmed.

For the farmer, the biggest change of the 1850s was the arrival of the railroad. The railroad ended the isolation of McLean County and fundamentally altered the corn farmer's life. With regular rail service about to start in May of 1853, one local newspaper said, "to the great body of travelers and emigrants the beautiful interior portion of the State has been a sealed book. But the throbbing of the great herds of the commercial world will henceforth send their pulsations into out midst" (*Bloomington Intellegencer*, 18 May 1853). Contents of the first arriving and departing freight cars were indicative of the nature of this change. The most important incoming freight was pine timber from Michigan and Wisconsin. Tons of inexpensive softwood flowed from the north into the county, touching off a major boom in the building of houses and barns. In addition, tens of thousands of board feet of planking were unloaded and hauled out onto the prairie for fence boards. Overnight, groves of local hardwood lost their economic importance; Northern pine was cheaper. Agricultural machinery could now be quickly ordered and delivered. Some of this was paid for by shipping livestock, but most came from thousands of freight cars loaded with corn.

Suddenly, McLean County had the attention of the nation. From all over the east came inquiries about the county. These inquiries were not always politely answered. When one potential migrant asked if the prairie grass was good pasture for sheep, the Intellegencer responded in print. "Yes sir, very good, and asses do extremely well on it" (*Bloomington Intellegencer*, 22 Feb. 1854).

Figure 2:14. The Dennis Kenyon homestead as it appeared in the 1874 Atlas of McLean County. Note the log cabin which survives in a clearly secondary role. (MCHS)

Figure 2:15. The Dennis Kenyon Home in the early 1970s. This is one of several New England housetypes found in the county. (W. Walters)

Figure 2:16. The construction of three railroads in the 1850s completely altered the relationship of the corn farmer with the world outside McLean County. (Illinois State Historical Library)

By 1860, McLean County's reputation was well established. One train traveler published an account of a railroad journey across Illinois from south to north. When the train reached the southern limits of McLean County, the conductor, rather than call out the county's name, informed the passengers that they were now in "God's own country." The commentator went on to remark, "such a burthen of grain! — such a promise of wealth! 'It grows better and better', said our French friend Perrault — 'it is very good', which was the highest commendation we heard him bestow on anything even when extravagantly enthusiastic. The emphasis was given, however, by industrious friction of the hands." The land was seen to be so rich that a traveler, and a French traveler at that, would rub his hands together in glee. High praise indeed ("The Canadian Excursion" *Prairie Farmer* 1860, 89).

There was a price to be paid for this sudden prosperity. Commercial success brought its own social problems. For all of its frontier roughness, there had been a rich social life in the timber-edge communities. Farmers and their families lived within daily talking distance of each other. Moving out onto the prairie greatly increased distance between farmsteads, reduced the possibilities for daily conversation, and often brought a sense of personal isolation. This apartness was a particular problem for the young farmers' wives. They were bound by children to homes that were no longer within easy walking distance of each other. In the 1850s the problem was not yet severe, but it was beginning to be noticed. In June of 1854, a perceptive reporter for the *Pantagraph* took the Illinois Central Railroad south, in the direction of Clinton. As he looked out at the newly settled prairie, he reflected, "These prairie farms are too large, and the dwellings too far apart, to give a character of sociability and good neighborhood. Scattered over the boundless land so sparsely, these houses of the sturdy yeomen seem more like half-way houses on an infrequently traveled country, than the neighborhood dwellings of the soil" (*Weekly Pantagraph*, 14 June 1854, 2). This observation is both valid and profound. During the second half of the century, increasing numbers of farm families would find Cornbelt farms to be half-way houses rather than lasting homes.

Still, by 1860, McLean County had reason to be pleased with the efforts of its farmers. Three-quarters of the land was officially listed as "improved." The remainder was mainly unbroken prairie, most of which was in the largely unsettled eastern townships. Unimproved land also included a substantial number of acres in woodland, much of this along the Mackinaw River. Statistically, the county had, in a remarkably short period, jumped to a position of national importance. There were over 11,000 horses and more than 50,000 hogs. The county produced huge amounts of oats to fuel the horses, an abundance of hay, and more than a million bushes of

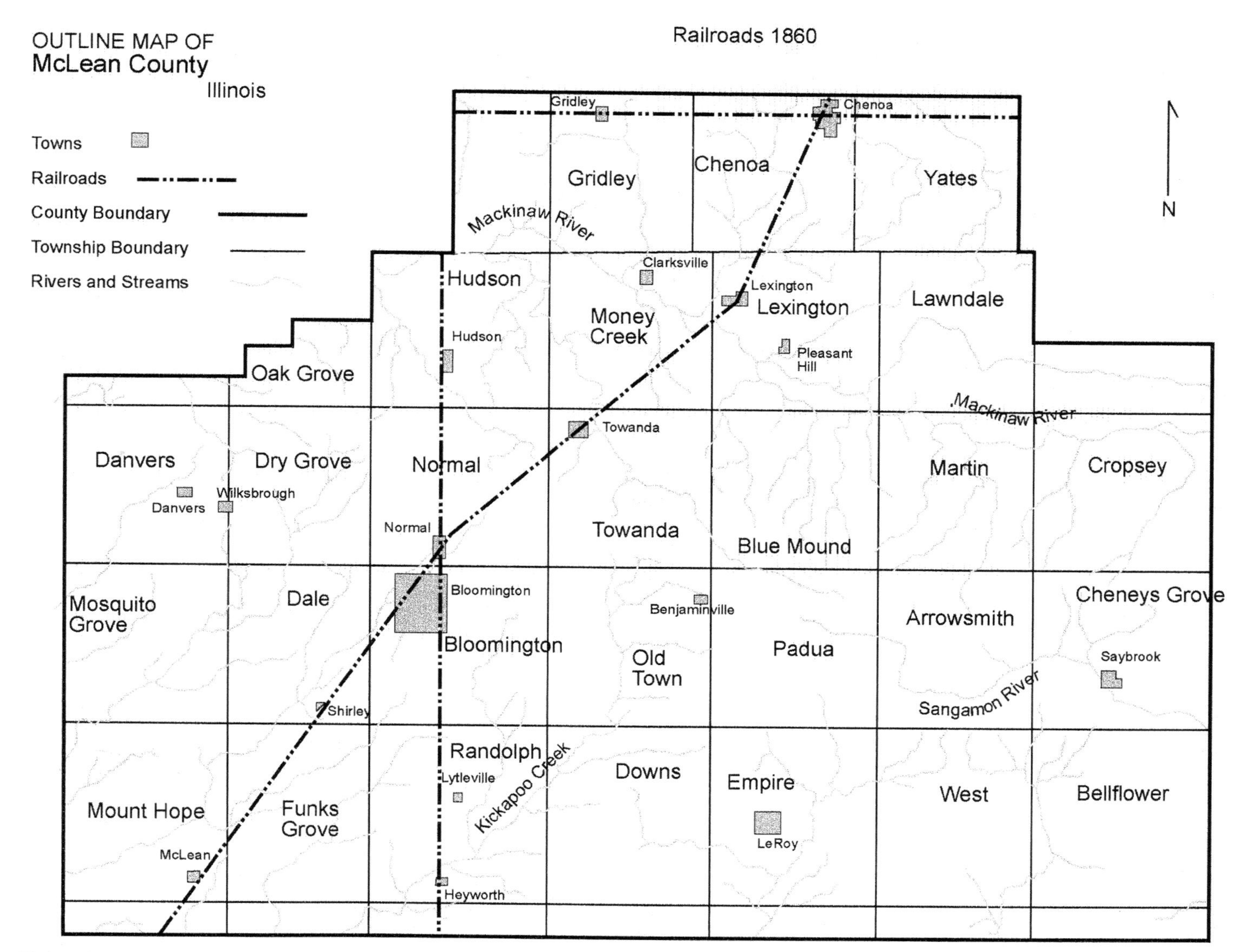

Figure 2:17. McLean County railroads in 1860. Note how large sections in the south and southeastern part of the county are still far from the nearest tracks. (MCHS)

wheat. These totals by themselves would have made almost any other county in the Union proud of its agriculture, but in McLean County, they were secondary. Where McLean County really excelled in was growing corn. Between 1850 and 1860, Illinois had passed Tennessee and Ohio to become the leading producer of Indian corn in the United States. The totals for 1860 were truly impressive. McLean County had exceeded all expectations and harvested more than three million bushels of corn. Skeptics might grumble that Sangamon County had produced slightly more corn than McLean and was the state's leading corn county, but Sangamon had a ten year head start on McLean. McLean County folks knew that they had better soil and better farmers. Everyone was certain that, come the next census, they would lead the nation in corn production. The year of 1860 was a time for bragging.

How this agricultural leadership was accomplished can only be understood by examining the way in which corn was produced in the middle of the nineteenth century. This can best be done, not by looking at statistics, but at farmers' lives. Fortunately, good records have survived and the agricultural year can be explored through the corn farmer's own words.

CHAPTER III

THE SEASONAL ROUND

By the time of the Civil War, most McLean County farms had passed the pioneer stage and had entered into a pattern which would remain essentially unchanged for the remainder of the century. While the economics of farm life changed greatly between 1860 and 1900, the essential yearly round of corn growing remained unaltered. Without understanding the seasonal round imposed by the dictatorship of the corn plant, no understanding of Midwestern life — from politics to courtship — has any real meaning. Reformers railed against its wastefulness, and inventors filled the shelves of the patent office with schemes to end backbreaking labor, but practical farmers recognized that the system of Cornbelt farming did indeed work. Perhaps the best way to follow this seasonal round is to look at the diaries of McLean County farmers during the last years of the Civil War.

Richard M. Britt was a twenty-six year old, Illinois-born, farmer living in the eastern part of the county. Completing Britt's household in the spring of 1865, were his pregnant, 24-year-old wife, Rhoda, whom he calls Med, and a teenage farmhand, George Wallace. The Britt farm was one of the most valuable in the township. Britt was a progressive farmer, and an avid reader of farm journals, including the *Prairie Farmer* (Britt diary).

The corn season began in spring as soon as the land was dry enough to work. Often this period could be the most frustrating time of year for the impatient farmer, and many filled the hours with a wide range of general farm activities. Perhaps the first activity directly related to raising corn was removing stalks left from the previous year's crops. At any rate, this is how Richard M. Britt began his corn year. On April 10, 1865, he went into the timber and cut a pole for a stalk rake. Four days later, the rake was finished and he began clearing corn stalks. The ritual of selecting new poles for stalk rakes seems to have been an important one; a number of farmers mention that activity in their dairies. Using such a rake was hard work. One farmer recorded, "went over to Don Hoovers to get stalk rake. Killing using it" (*Jessee* 7-31). Green corn stalks had a certain value as fodder but brown stalks from the previous year's crop were simply a nuisance. They were generally raked into piles for burning or plowed under. Raking stalks was not very remunerative work, but it served to pass the time in a useful way while waited for the land to dry and for the danger of spring frost to pass.

The start of planting was often the most important date in the farmer's year, but it was preceded by a number of field preparation activities, and the first of these was plowing. For the farmer waiting to plow, early spring was a time of seemingly endless anticipation. Each farmer watched the sky, walked the damp fields, and above all, carefully watched his neighbors. The ritual of watching neighboring farmers has always been one of the great constants in McLean County corn farming. What crops were they going to plant? What new machinery were they buying? Above all, when was he able to start work in the field each spring? Obsession with a neighbor's crops could reach near pathological proportions and has long been a cause of marital stress.

Plowing was the first major field activity, and spring plowing was certainly the rule in nineteenth century McLean County. On April 19,1865, Richard Britt noted anxiously that one of his neighbors had begun plowing for corn, but his own fields were still too wet. Impatiently, Britt waited, filling his time reading *Coleman's Rural World*, cutting and twisting hedge, or burning grass in the slough. Early planting gave the farmer a potentially longer growing season, provided more time for possible replanting, and permitted more flexibility to work around bad weather. Until the 1880s there was no economically practical system of under-drainage and farmers could do nothing to speed the work of drying their fields.

May 1 came and Britt recorded glumly that it was still "too wet to farm." On May 3, he went into Atlanta to buy a new corn planter and family groceries. The roads were deep with mud and nearly impassable. That morning, Lincoln's funeral train passed a few miles south of Britt's farm, but Britt noted that he had missed seeing it. The big day for Richard Britt came on May 4 when he at last was able to write, "Plowed for corn." On the next day, he scoured his Indiana breaking plow and, in spite of the wet ground, got some plowing done. But Britt was well behind; a neighbor had already started planting corn. By May 9, Britt recorded that he had not quite finished plowing but "was so sore after yesterday's work can hardly walk." On May 12, plowing was at last done and Britt could begin the next stages in field preparation: harrowing, rolling and marking.

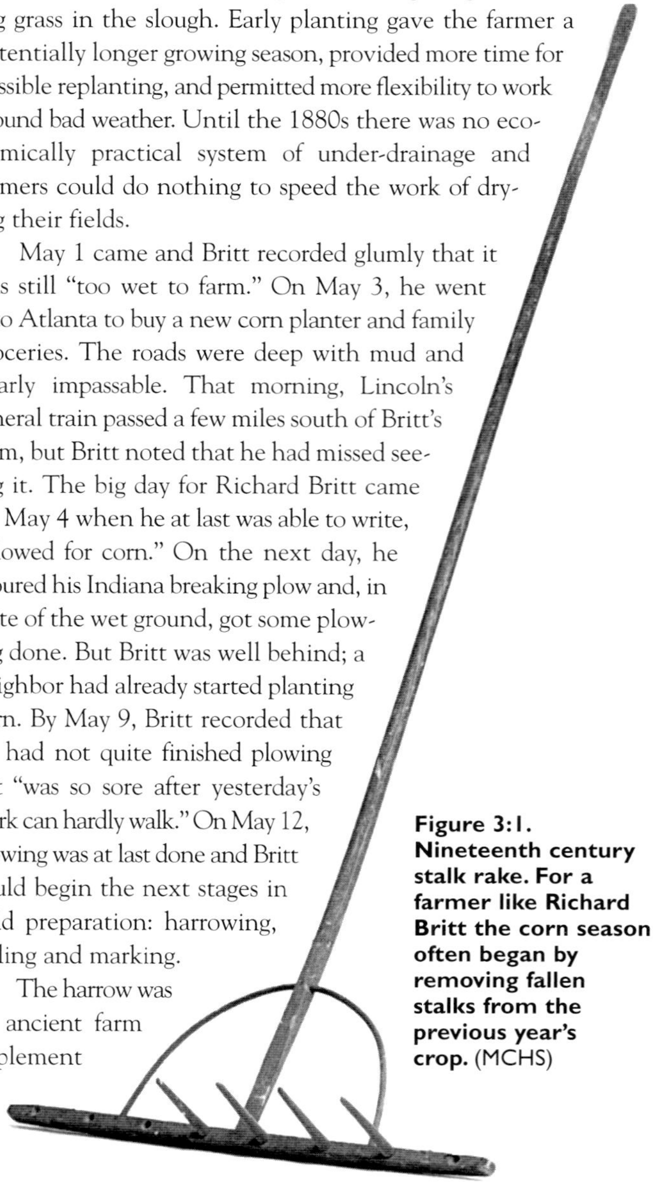

Figure 3:1. Nineteenth century stalk rake. For a farmer like Richard Britt the corn season often began by removing fallen stalks from the previous year's crop. (MCHS)

The harrow was an ancient farm implement

Figure 3:2. Cultivation was essential to contol weeds. (*Prairie Farmer*)

which, in its simplest form, consisted of nothing more than a spiked board drawn behind a team of horses. However, by 1865, a number of factory-made harrows were available. The function of the harrow in field preparation was to level the ridges left by plowing and to break up clods of earth, thereby permitting corn roots to penetrate the soil more easily After planting, harrows might also be used to cultivate between rows. Large rollers were then pulled across the fields, often drawn by a separate team working behind the man with the harrow to further pulverize the field. The ideal was velvet smoothness before planting. Next the farmer "marked" the field into a grid of squares to indicate where the corn should be planted. Marking might be either a walking or a riding activity. Britt's marker seems to have been horse drawn. At any rate, it had a seat which needed repair at the end of the planting season. Markers might also be very simple, and in the 1860s homemade corn markers were still popular.

Seed selection was always done with care, but it remained — at least by modern standards — extremely unscientific. At harvest, the farmer simply selected the best looking ears and tossed them into a separate bin to use for the next year's crop, or perhaps hung them on a metal drying rack. Seed corn lasted longer if it was stored as ears and was not shelled until planting time. Already, however, the rudiments of a commercial seed corn industry existed. For example, on March 29, 1860, Osborne Barnard, a farmer living just west of Bloomington, recorded that he had, on that day, shipped two barrels of seed corn on the Illinois Central Railroad to a man named J. A. G. Inskeep (Barnard Diary).

Most farmers , however, shelled their own corn, either using any one of a number of mechanical shellers or by the ancient expedient of rubbing one ear against another. On May 11, 1865, Richard Britt began shelling seed corn and then went over to a neighbor's to see about getting a planter. Perhaps it was his own Brown's Corn Planter

Figure 3:3. Like many farmers Britt used a rollers to break up clods and help level the field. (*Rural Affairs*)

Figure 3:4. Marking the cornfield was essential. If plants were not evenly spaced they could not be properly cultivated. Note the hand held planter. (*Prairie Farmer*)

which he had purchased two weeks earlier. Indeed, nineteenth century farming would have been an impossibility for all but the wealthiest were it not for the continual borrowing and lending of labor, teams, and machinery. To comprehend the social fabric of the rural Midwest demands that one understand how it was continually reinforced by the trust implicit in the system of lending needed items. Throughout the spring of 1865, Britt's farming tools regularly vanished and reappeared as they were needed by his neighbors or himself.

There was folklore concerning the proper date for planting corn. Some said that a wise man waited for the day when he saw the first apple blossom, while others maintained the proper indicator was the first sight of green leaves on the oak tree. Whatever indicators he used, Richard Britt decided that May 12 was the proper

Figure 3:2. Britt periodically shelled stored ears to sell or provide feed while he waited for the new crop to mature. There were many sorts of corn shellers. (*Prairie Farmer*)

date. On this day, he made a tentative start by planting two small patches of corn and by shelling seed for additional planting, but then it once again began to rain After the delay, planting began in earnest, and for the remainder of May, Britt, together with his hired man, George, continued to plow and to plant corn. The Brown's Corn Planter purchased by Britt was a two seat machine which had first been patented in 1855. The driver steered the apparatus while a second person, often a boy, was seated crosswise, with the job of pulling a lever whenever the planter crossed one of the intersections of the marked grid and depositing several seeds into the hill.

While a farmer might speak of the "planting season," it is clear that planting corn could extend over several weeks and was always interspersed with other activities. It is worth noting, for example, that before his own corn was fully planted, Britt took time to help a neighbor harrow and mark. Shelling corn for seed continued throughout the planting season. The last day of May found Richard and his wife, Med, now in her ninth month of pregnancy, in the barn shelling more seed corn.

The dilemma over corn varieties is reflected by Britt's notes for June 2, 1865. He and his hired man, Charles, began planting the field in front of Britt's house. After planting about eight acres, Britt became concerned. He stopped work, took up the seed, which he describes simply as "yellow corn," and examined it carefully. After the inspection, he decided that the seed corn was of poor quality and called a halt to the planting operation. Returning to the barn, he found some white corn which looked to be of better quality, began shelling, and by noon was back in the field. Richard Britt was not an emotional man, and his written words are rarely emphasized, but on the following day, he did underline three words in his diary, "we managed to do a big day's work and make a finish of planting corn for this year."

What Britt meant was that he had made a finish of the orig-

Figure 3:3. Brown's Corn Planter. Britt purchased a machine like this in the spring of 1865. He was an avid reader of agricultural journals including the *Prairie Farmer*. (*Prairie Farmer*)

Figure 3:7. R.M. Britt, Mt. Hope Township farmer. (P&B)

inal planting of corn. Nineteenth century McLean County farmers all expected to do a certain amount of replanting. Water was both the great friend and the great enemy of Cornbelt farming. In most years, standing water and washouts near the numerous shallow waterways — universally called "sloughs" — would destroy hundreds of acres of young corn plants. It was still May when Britt found much corn "badly missing" along his slough. Replanting was the only answer and replanting corn was always difficult work. One farmer called it "slow work. Very tedious job." At the same time some corn was replanted, cultivating corn in other fields had already begun. Field work was primarily a male activity, but everyone was busy, including Britt's wife, who had been ill, but by June 4 was feeling well enough to feed her chickens. A week and a half later she gave birth to her first child, a girl. Richard Britt recorded with characteristic gruffness that the babe "was not yet worth a name." Eventually, the Britts decided to call her Hallia.

In a year like 1865, when corn was slow to mature, cultivation was critical. By the 1860s, much corn cultivation in McLean County was done by machinery. Farmers would certainly have agreed with the Midwest's most famous agricultural writer, Solon Robinson, who remarked that "instead of hand hoeing, use some of the light horse hoes, with which one man will do more good than ten with hand hoes, following a mold-board plow" (Robinson 1867, 726). The tools of cultivation were often the same sorts of plows and harrows used in preparation, which could be adjusted for the width of rows but riding cultivators, like the one owned by Britt, were becoming increasingly popular. Britt recorded on June 6, "I roll the corn and George harrows." Alternately, he and George might use the plow and harrow together. Both riding and walking plows were in use. On June 19, Britt recorded, "George and I plow. He with the single shovel and I with the riding plow." By single shovel, Britt meant a plow without a mold-board that broke the earth and cut the roots of weeds without massively turning over the sod. On June 23, George broke the shovel plow and Britt had to take it to Atlanta for repair. Whatever the tool, the essential purpose of cultivation was weed control, and this tilling continued on Britt's farm throughout the month of June. By July, the plants were high enough to shade most weeds, and the corn was left to fend for itself until the middle of September and the beginning of harvest.

Farm animals were the most easily overlooked members of the team that created the great American Cornbelt. For all the much publicized improvements in seed and farm machinery, the lot of animals — especially draft animals — remained demanding and not infrequently, cruel. The reality of prairie farming was that clay soils were difficult to work. When it was dry, they could bake into rocklike hardness. When it rained for several days, they held water and became heavy, fouled harrows, clung to plows, and sometimes could elicit from even the most Christian of farmers some very harsh language. Harnessed beasts of course lacked even this means of reflecting their pain and frustration. Moreover, as the century progressed, farms became larger, and, for the farm animal, much-heralded improvements in machinery often meant that the

Figure 3:8. Britt horse Rosey created problems for the farmer. Here a contemporary advertiser suggests an answer for his problems. (*Prairie Farmer*)

Figure 3:9. Corn was the object of Britt's efforts. By 1860 McLean county was already one of the leading corn producing counties in the nation. (Galinat)

hauling of and greater weights of iron and steel through the sticky clay. Events in the spring of 1865 on Richard Britt's farm in Mt. Hope Township suggest something of the animal's life.

As noted earlier, Richard Britt began plowing on April 29, 1865. Even though he had borrowed an additional team from neighbors to help with the work, it was brutally hard work for the animals. By June 2, he was forced to record that "the horses were almost worked down." Three days later, near the end of a long day's planting, one of Britt's horses dropped in the harness, and Britt reported that all of the horses were unfit to work. He ascribes their poor condition to a combination of bad hay and hard work. "Good hay, or none," he resolved, was henceforth going to be his rule. On the next day, by the expedient of alternating teams, he finished planting his corn, but the horses still had tough days ahead. On June 7, 1865, Britt was cultivating corn with the Leeper and Kidder riding plow which he had purchased in April from Dills and Howser in Atlanta for $65.50. However, for the horses, this new kind of riding cultivator simply added the farmer's weight to the burden they were already pulling. One of the horses, Messenger, pulled without protest, but the other, Rosey kept, refusing to work. Britt finally became so angry that he burst into Biblical verb forms and recorded, "She baulks so that she causeth a storm of wrath and I thrash her furiously."

The most anonymous members of the corn-raising team were the farm laborers. Unless enumerated by the census taker, we often know only the first names of this essential group. For example, we know that Richard Britt's hired hand was George Wallace, and that he seems to have been a full time employee. Britt also wrote of Celly, Mat, and Charles working for him at various times in the

Figure 3:10. James Wilson Jessee. When he returned from the Civil War Jessee worked for a number of McLean County farmers and still found time for church, courting, brawling, and politics. (W. LaBounty)

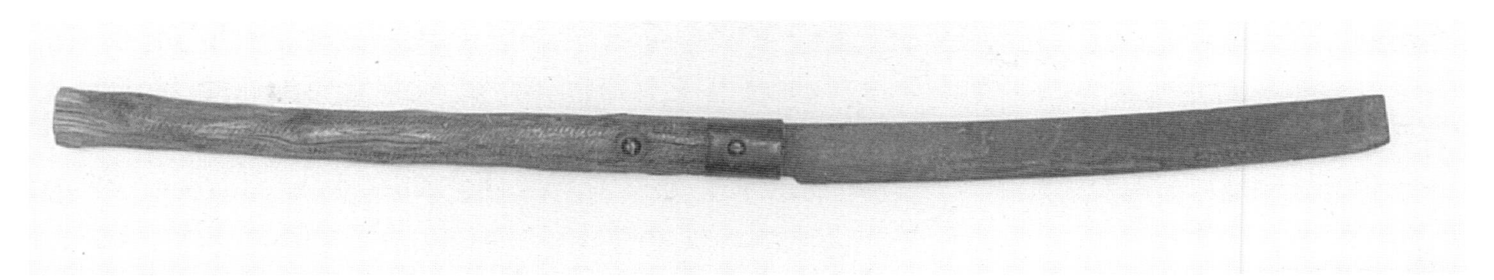

Figure 3:11. The corn knife was one of the most widely used tools in McLean County. They were often fashioned from pieces of scrap metal. (MCHS)

spring and early summer of 1865, but we cannot be sure of any of their last names. Some were probably relatives, sons of neighboring farmers or even the neighboring farmers themselves.

Some farm hands were clearly transients. On June 12, 1860, Osborne Barnard hired "French Charlie" for 13 dollars a month. On September 3 of that year, a "stranger" knocked on Barnard's door and asked to stay the night. The next day, Barnard and the "stranger" worked together in the field, tying up shocks. Such temporary labor was often unreliable. More than fifty years after the event, Samuel Baldridge recalled with considerable bitterness the time he hired a "jail bird" (Baldridge Memoirs). Often neighbors would band together to form large harvesting crews which moved from farm to farm.

In slack seasons, unemployed farm laborers wandered into the local towns. Here they formed a class of people known as "the boys," who greatly enlivened small town life. They rounded out baseball teams and added their voices to political rallies. They sat on benches, drank whiskey, and shouted obscenities at local women who lifted their skirts when trying to cross muddy streets. No one disputed the fact that they were economically essential, but town merchants always breathed a collective sigh of relief when harvest or planting season arrived and the boys once again drifted off into the corn fields.

In July and August there was relatively little work in the corn fields. The corn was too high to do much cultivation, and none of the various insect control measures available was of much practical value. There was, however, much checking of fences. The large wandering hog population devastated many acres of corn, and cattle loved ripening ears. By late summer, McLean County farmers were looking for signs that ears were mature. When this happened, the leaves begin to turn brown; the farmer would begin to peel back selected ears searching for black at the base of the ear. A blackened base indicated the corn plant had completed its work and nothing more would add to its nutritional value. At the same time, the surface of individual kernels took on a glazed, hardened appearance. Of course, part of the farmer's art was guessing from the appearance of a few ears when an entire field was ready for harvest. Cutting corn began in September or October. Unfortunately, Richard Britt's carefully detailed diary does not extend into the harvest season, but a typical season's activities can be followed from the comments made by another McLean County farmer.

In the fall of 1864, twenty-six year old James Wilson Jessee had recently been mustered out of service in the 8th regiment of the Illinois Volunteer Infantry. Like many ex-service men, he had

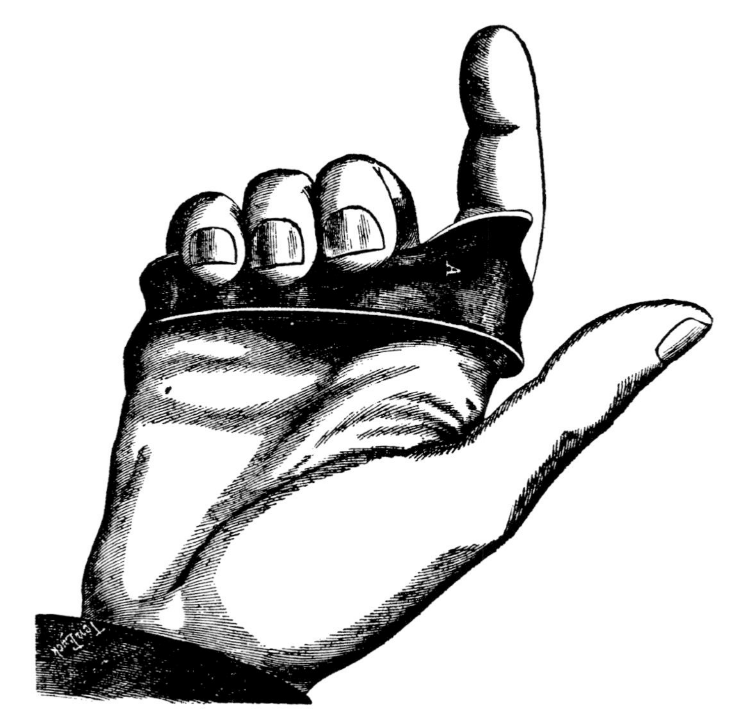

Figure 3:12. Every farmer in McLean County was familiar with husking hooks. They were widely used from the earliest settlement until the 1930's. (*Prairie Farmer*)

a little money in his pocket, was uncertain about his future, and was in no hurry to slip back into the routine of a working life. His diary records odd jobs, lots of preaching and politics, much socializing, buying new clothing, and generally enjoying himself. When he heard news of a relative's marriage, his diary summed up his general feelings with the words, "Plenty of time for old Jim yet." Jessee was an interesting person who could plant cabbage in the morning, go to church in the evening and, between the two activities, record that he "had fuss with Jake Vanderventer. Rebel deserter. Choked him a little and gave him a lecture." His great hatred was Copperheads; that is, northerners who favored ending the war. Jessee used the term Copperhead — which he usually abbreviated as cops — to mean anyone who advocated voting for McClellan and he used "cop" and "traitor" interchangeably (Jessee Diary).

Politics and preaching still left time for more mundane activities and corn harvesting called for all available hands. Beginning on Wednesday, September 13, 1864, James Jessee recorded his participation in the McLean County corn harvest. "In good Health as common.

Smaller varieties of corn, in smaller hills, will enable the expert laborer to take six hills at a time, and to form a large shock of 49 hills, (fig. 38;) commencing at *a*, he takes the first three as a beginning; next at *b*, he takes six; at *c*, the next six, and so on, the dotted lines showing his footsteps. A larger armfull may be taken by placing the arm above and before the hill, instead of behind it.

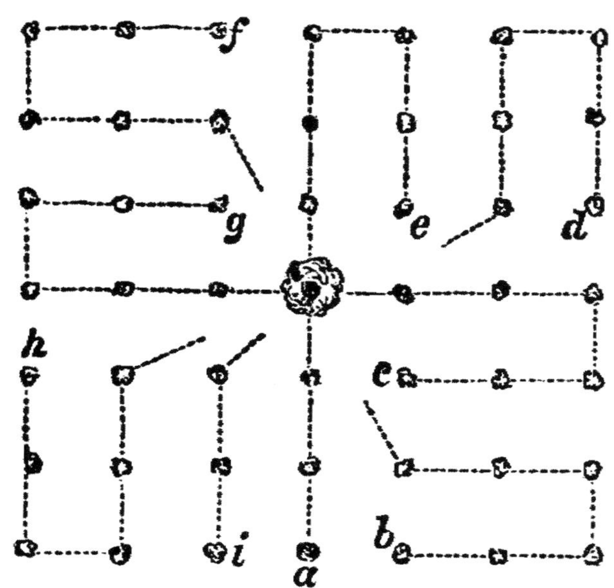

Fig. 38.—Mode of Cutting up Corn, forming a Shock of 49 hills.

The common mode in cutting is to place the shock around a central uncut hill, which occasions some inconvenience in husking, to obviate which the corn horse is used. It consists of a pole about 12 feet in length, and nearly as large as a common wagon tongue. One mode of constructing it, (shown in fig. 39,) is by placing the legs at the end of the pole, the other end resting on the ground. Two or three feet back of the legs a horizontal hole is bored, admitting loosely a rod four or five feet long. The corn when cut is placed in the four corners made by the rod and pole, and when the shock is finished, the rod is pulled out and the pole drawn backward. In fig. 40 the same end is accomplished, only the pole is drawn forward instead of backward.

Fig. 39.—Corn Horse, used in constructing Shocks.

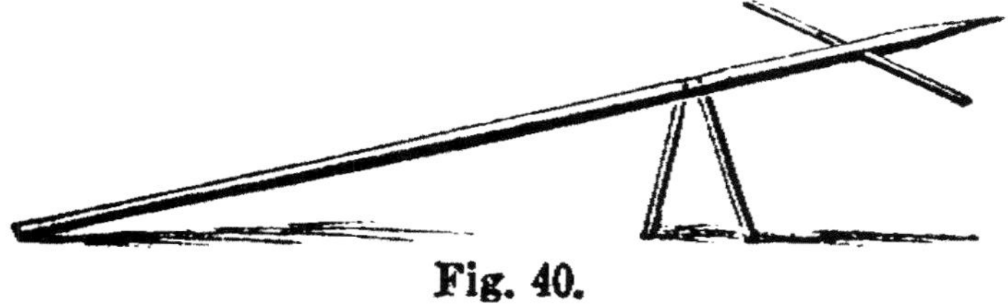

Fig. 40.

Figure 3:13. The work of putting corn up in shocks is described above. (*Rural Affairs*)

Up Early. Weather clear But cool and pleasant. Commenced cutting corn this A.M. for Holloway at 8cts per Shock twelve hills square worked hard all day. Cut 35 shocks. Corn heavy and a little green made $2.60 think will get rich at the Business. Times very good generally." And for the following day: " Cut corn again cut 35 shocks. Was right tired when night came. Was Bothered some. Worked till dark, hard work goes pretty tough." And the next day: "cut corn again." By Saturday, Jessee was "quite sore" but still managed to stay up much of the night with a borrowed revolver, guarding a flagpole which the local Lincoln forces had erected as part of their presidential campaign.

The essential harvest tool of the mid nineteenth century was still the simple corn knife. This implement was slightly curved, about 18 inches long and was often hammered out from any available surplus bit of metal. Although mature, the moisture content of the ear at this stage was too high for storage. The problem which

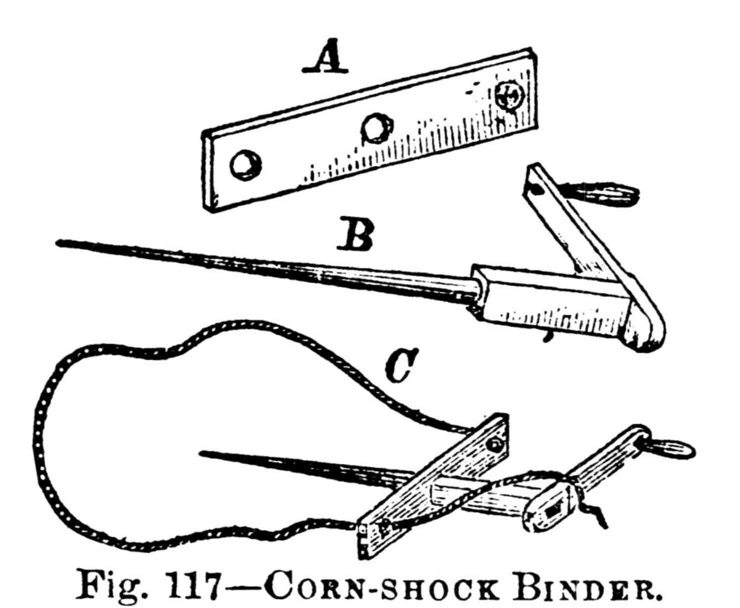

Figure 3:14. Corn-shock binder. (*Rural Affairs*)

the farmer faced at harvest was to encourage drying while still allowing for eventual efficient separation of ear and stalk. The practical answer to this was the shock.

To create a shock, corn was cut at ground level and the stalks with the ears still in place, were bundled in groups. These clusters of corn stalks were begun with the aid of a light frame of sticks, which was removed when enough of the shock was in place to permit it to stand by itself. The stalks in a shock were bound near the top. Traditional binding materials included wilted suckers, small corn stalks, or rye straw, but by the 1860s, twine was becoming increasingly common. The shock was left in the field until the ears were completely dry. To understanding the need for shocking, one must keep in mind that nineteenth century corn varieties lacked the standability which is now bred into the corn plant. Therefore, if left standing by itself in the field, the corn which Jim Jessee was harvesting was far more likely than twentieth century hybrids to topple and leave the ears on the ground where they would quickly rot.

Following shocking, the next step was to separate the ear from the cornstalk. The shock made this operation much easier by

Figure 3:14a. Shocked Corn. (*Pantagraph*)

Figure 3:15. Corncrib (upper right) on the Lewis Bohrer farm shown in the 1874 *Atlas of McLean County*. (MCHS)

keeping the corn in an upright position. The lower end of the ear was grasped firmly between thumb and fingers of one hand, and the other hand was used to force the ear over the back of the hand which was grasping the corn. This process could be done without disturbing the binding which held the shock together. The ear, still in its covering sheath, was then tossed into a wagon. When filled, the wagon was returned to the farmstead where husking could begin. Husking involved the manual stripping of the ear from its enclosing sheath of leaves and was done with a glovelike canvas device fitted with a sharp steel hook. Competition for labor at harvest was intense and farmers competed with each other to make husking an attractive social event.

Jim Jessee's first mention of husking does not occur until November of 1864. After a "jollification" meeting to celebrate Lincoln's victory, at which Jessee spoke for about an hour, giving the Copperheads "fits." Jessee writes that he has declined an invitation to a corn husking at Mr. Gordon's. However, on November 23, he records that he has been husking at Woodcock's. On November 29, he has another invitation, for a husking and oyster supper at Bob Ball's. Again, he must refuse because he is working on a threshing crew, but he does manage to show up for the oysters. Unfortunately, the evening is spoiled by the large number of "cops" present. A third invitation comes on December 5 for what Jessee calls, with tongue in cheek, "dinner and supper with Mr. Keerans for which they only ask me to Husk corn all day." This time he decides to attend and spends Tuesday, December 6, at a husking and quilting affair. On the following Thursday, a husking and turkey dinner is scheduled at Craig's, but it is too cold to husk and no one else comes, so he spends his time helping Craig haul wood. On Friday, he returns to Craig's and they spend the morning husking.

For Jim Jessee, corn husking continued into the new year. On December 16, 1864, he hitched up his newly purchased team of colts and began gathering corn. He writes that he has husked two loads for Aunt Margaret and then "jerked a load for ourselves." On the twentieth, he husks all day. In February, in "fine weather warm and pleasant," he again spends the day husking corn.

Once the ear corn was husked, it was usually stored in rail cribs. Easterners often regarded such warehousing of ear corn as slovenly farming, but writers like Solon Robinson, who knew the Midwest, regarded rail cribs as both efficient and essential. They were built from standard length fence rails, with half of them then cut into equal lengths. The rails were then notched and built up into an oblong structure half as long as it was wide and about eight feet high. The roof of the rail pen was covered with clapboards or straw and, wherever the ears threatened to spill out, straw or board chinking was added. Hundreds of these structures were built in McLean County and they were not necessarily temporary. Part of a pioneer rail crib was still in use on the Benjamin farm, in eastern

McLean County, in the late twentieth century. Not the least of their advantages was the speed with which they could be erected. On November 24, 1864, Jim Jessee recorded that he had spent the afternoon hauling rails and then building a "pen" to hold the freshly husked corn.

Not all corn was stored in rail cribs. Some prosperous farmers built wooden cribs about 30 feet long and 20 feet wide. They were frame or heavy timber structures about ten feet high with distinctive central passages wide enough to permit a wagon to be driven inside and unloaded. Above the passage, there was often a fully walled area for the storage of small grain or shelled corn. On either side of the passage were slatted cribs. Once inside, the farmer could throw corn over the beam either to the right or to the left. Slat or lattice work gates in the end permitted free circulation of air through the bin. Oak, which was available from local groves and extremely durable, was a popular wood for such cribs. Moreover, it could withstand considerable stress, which was extremely important; as ears of corn settle, they exert considerable pressure on the structure and sides of the crib. As long as corn was stored as ears, this kind of crib was a vital part of the Cornbelt economy as well as a distinctive landscape feature. Many McLean County cribs are substantial buildings. Because of their size, nonfarmers frequently mistake them for barns.

Neither rail nor board cribs were entirely satisfactory and much effort was made, during the next hundred years to improve their design. To keep air circulating around the ears without exposing them to the dampness of rain and snow was the problem and it never really solved until the advent of combines in the 1950s made any sort of structure for the storage of ear corn redundant.

Once the corn was harvested, the farmer had a number of options. One option was to sell the corn "in the ear." In the summer of 1865, Richard Britt was getting from 30 to 35 cents a bushel from the sale of the previous year's corn. Although figures varied widely, an expectation of 35 cents was not unduly optimistic. Alternately, the farmer could shell the corn himself — or have it shelled at the local mill — and market the corn as grain. Grain, of course, commanded a higher price per bushel, but shelled corn could not be stored for as long as ear corn before it "went bad," so shelling had to take place soon before the date of anticipated sale. Many farmers insisted that the most economical use of corn was to grind it, ear and all, to feed to stock as meal.

There were secondary sources of income from corn. Stalks had a certain market value; one Illinois farmer estimated, in 1852, they were worth about $1.50 an acre. Harvesters always missed a substantial number of ears which were gleaned by cattle and hogs turned into the fields after the shocks had been cleared. Moreover, even in the best years for corn, many other crops were produced. Large amounts of land were devoted to wheat, oats, and hay. Most farmers had substantial fields devoted to special crops, including potatoes, a wide range of vegetables, and a surprising amount of fruit. Cattle and hogs were major sources of cash for nearly everyone. Horses were bred in large numbers. Sheep were popular and chickens nearly universal. Unimproved land, including timber, honey, game, and prairie hay, provided some income. For the majority of McLean County farmers, however, nothing proved as valuable as corn, and the seasonal round continued essentially unchanged for seventy-five more years.

CHAPTER IV

DRAINAGE AND DISCONTENT
1870-1890

In McLean County, the late nineteenth century was a period of both triumph and discontent. The triumph was evident in all production figures for corn. In 1860, McLean County produced an astonishing 3,228,960 bushels of corn. By 1870, that figure had jumped to 3,723,379 bushels and there was better to come. Between 1860 and 1880, the county witnessed a great surge of building. Great numbers of new barns were erected, corn cribs were expanded, hundreds of new farmhouses were constructed, and the county quickly lost most of its pioneer appearance. Pre-Civil War vernacular architecture, including log cabins, is scarce in rural McLean County because so much of it was replaced by the great prosperity of the late nineteenth century. Farmers purchased an astonishing amount of new machinery, and the impact of their purchases was felt in dozens of small towns. The farm equipment dealer, with direct telegraphic links to Chicago, was rapidly supplementing the local blacksmith.

An expansive mood seized McLean County. County farmers belonged to a generation accustomed to allegorical expression, and when politicians spoke of McLean County crowned with the laurel wreath of victory and presented with the cornucopia of plenty, they accepted such language as both accurate and just. When he addressed Illinois farmers in 1887, Adlai E. Stevenson said of McLean County that, "Its history stretching over but little more than half a century savors more of romance than reality." And he went on to proclaim, "I know of no people who have greater cause for gratitude to the Dispenser of all good than those I now address. Truly your lives have fallen unto you in pleasant places" (Stevenson 1887, 1-2). It was heady stuff, but no doubt many of Stevenson's listeners believed these words were true. Indeed, the accomplishments of the people of Illinois were amazing.

For the farmer, the greatest of these accomplishments were drainage and transportation. Tile drainage totally transformed the Cornbelt landscape. One of the largely unchronicled struggles of the McLean County corn farmer has always been the bitter war against standing water. In a way, it was an ironic struggle because it was this same standing water which created the black prairie soil on which the corn thrived. Of all soil-forming materials, clay particles are the finest, and to each particle clings an envelope of water. The result is a dense mass which acts to prevent the downward movement of surface water. After a heavy rainfall, much of the county was covered with a series of shallow ponds. Trees could not take root in such wet environments, but Tall Grass Prairie thrived. Each fall, the grass died and the blackened remains of its blades and stems were incorporated into the prairie soil. Each acre of prime prairie soil contains 150 tons of organic material (Cronon 1991 98). Moreover, unlike some soils high in organic material, prairie soils are not excessively acidic. North of McLean County is a vast arch-like rock formation known as the LaSalle Anticline. When glaciers passed over the anticline, they ground off thousands of tons of limestone from the top of the formation and incorporated these fragments into the material which they were carrying. Ultimately, this material was deposited in McLean County. Limestone fragments from the anticline help to control soil acidity. In short, the county possessed nearly perfect corn growing soil, but before drainage,

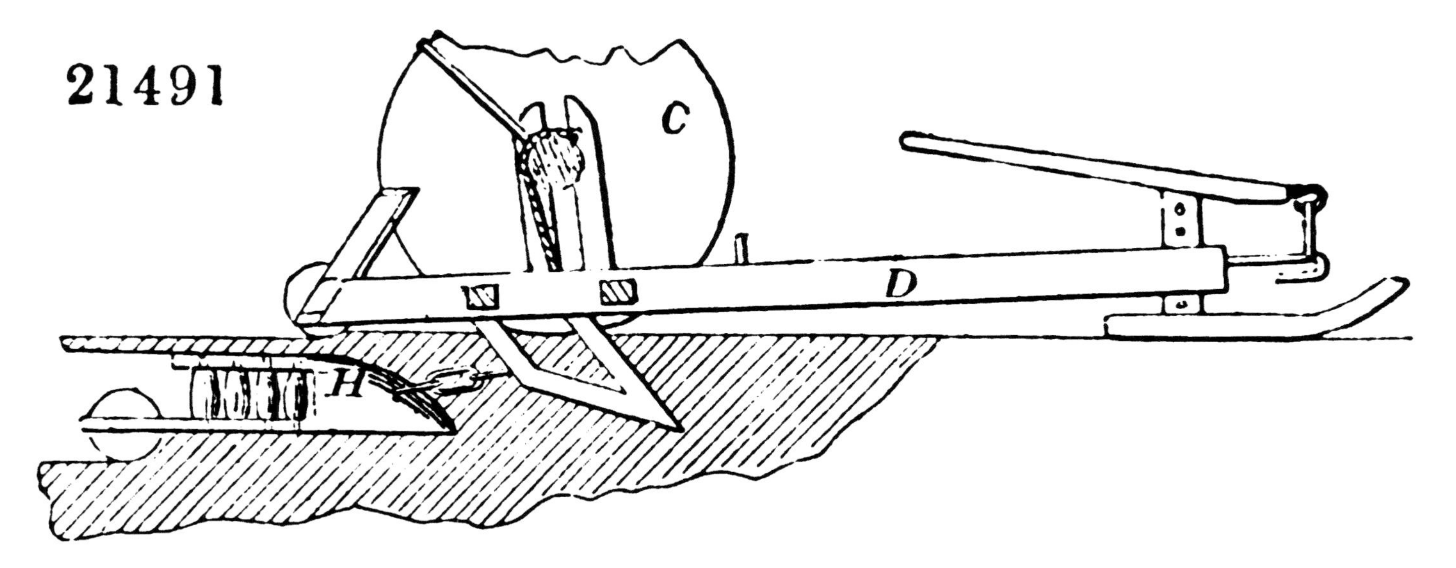

Figure 4:1. A mole ditcher. These were widely advertised but not very successful devices for draining farmland. (U.S. Patent Office)

4:2. The Heafer Tile Company at the turn-of-the-century when it was one of the most successful firms in Bloomington. (MCHS)

much of it was too wet to farm.

The history of land drainage is often a history of bad ideas. One of the worst such ideas was an ungainly contraption called the mole ditcher. The 1860 *Illinois Business Directory* included an advertisement for this device. Essentially, the mole ditcher was nothing more than an iron-tipped log, series of linked logs, or round iron blade about four inches in diameter which was pulled through the soil a few inches under the surface. In theory, this created an underground passageway through which excess water would drain. In practice, the mole ditcher was extremely difficult to drag through the ground. The usual method was to attach oxen to a long wooden arm affixed to a windlass. The windlass was linked by ropes or chains to a surface projection of the device. Problems were many. It was extremely hard to maintain constant depth. This was a major problem because successful drainage requires channels to follow a carefully regulated slope. Manufacturers of mole ditchers advertised that the small tunnel was efficient and long-lasting. Farmers found otherwise. Channels frequently collapsed, permitting water to back up which flooded the young corn. Roots choked passageways. Burrowing animals used the passages for highways and quickly altered their geometric patterns to suit their own domestic needs. Once blocked, mole-ditcher channels were difficult to clear. It is testimony to the McLean County farmer's severe problems with standing water that mole ditchers were used at all.

4:3. Early drain tile was usually two to four inches in diameter and had flat distinctive bottoms. Size soon increased and tile became round. (MCHS)

Much advice was available on drainage problems. In 1871, the *Pantagraph* copied a lengthy article by a Professor Schaltuck, who explained that two inch drain tile laid four feet beneath the surface was ideal for prairie soil. It looked fine in print. Local farmers were, of course, aware of tile drainage which had been extensively used in Europe for a generation; drain tile had been advertised in McLean County at least as early as 1867. All this was fine, but practical farmers also knew that drain tile was expensive and that transporting it any distance was sure to lead to extensive breakage. At that time, drain tile was made by the same process, and used the

4:4. Tile being laid about 1940. These tile are larger in size but are placed in the ground exactly as they would have been in 1880. The hidden secret to McLean County corn growing is the vast underground network of these tile. (MCHS)

TILE WORKS OF BARTELS & STOLL – LEXINGTON, ILLS.

4:5. Many country tile factories like this were in business for only a few years. (P & B)

same material that was used to make common brick. Like such brick, it fractured easily. The answer to the problems of cost and breakage was to manufacture the tile as close as possible to the farm, but this was easier said than done.

Edgar M. Heafer was the first to take up the challenge. He was a Bloomington native, an ambitious young man who would eventually become the city's mayor, but his father, Napoleon Heafer, was a hard-as-nails old-time brickmaker with a gruff pioneer sense of fun. Napoleon's idea of a good joke would enter into the story of Edgar's first attempt to manufacture tile. Young Heafer began his tile-making experiments with a used Penfield Press, which he purchased from a failed firm in Scott County, Illinois. The machine, which cost $375, was, in reality, nothing more than a large auger which squeezed wet clay through a die. A rotating wire then cut the toothpaste-like stream of raw drain tile into pre-set lengths, which were then dried and fired like brick. At least that was the theory. Young Edgar carefully supervised the transportation of the used tile machine to Bloomington. After days of work, the Penfield Press was at last attached to a long wooden sweep which was pulled in circles by a patient horse. Everything was made ready to produce Bloomington's first agricultural drain tile. Relatives and brickyard workers gathered to watch the event. Pugged clay was fed into the hopper on the press. The horse was clucked into motion. The auger turned. Everyone waited for the round-topped, flat-bottomed tile to emerge from the machine. Edgar Heafer was ready for the start of a new age in Cornbelt farming. To Edgar's horror, what came out of the machine was not carefully molded tile, but a shredded gray mass with the texture of damp soapflakes and no discernible form.

Edgar was stunned. He watched as his dreams of local fame and prosperity crumbled and he no doubt wondered how he would ever pay off the money borrowed to purchase the clearly-flawed machine. Eventually, his father could no longer hold back his laughter. Edgar realized he had been duped. The elder Heafer had secretly arrived early and mixed rough coke screenings into his son's carefully prepared clay. Later, Napoleon explained it was his way of teaching his son to always check his clay, but the father's actions were typical of an age which found such practical jokes hugely funny.

RESIDENCE, MILL & TILE WORKS OF JOSEPH DORLAND, SEC. 21, RANDOLPH TP.

4:6. In 1887 the steam engine for this small tile factory did double duty by serving Dorland's sawmill in winter. Note the down-draft kilns for firing tile. (P & B)

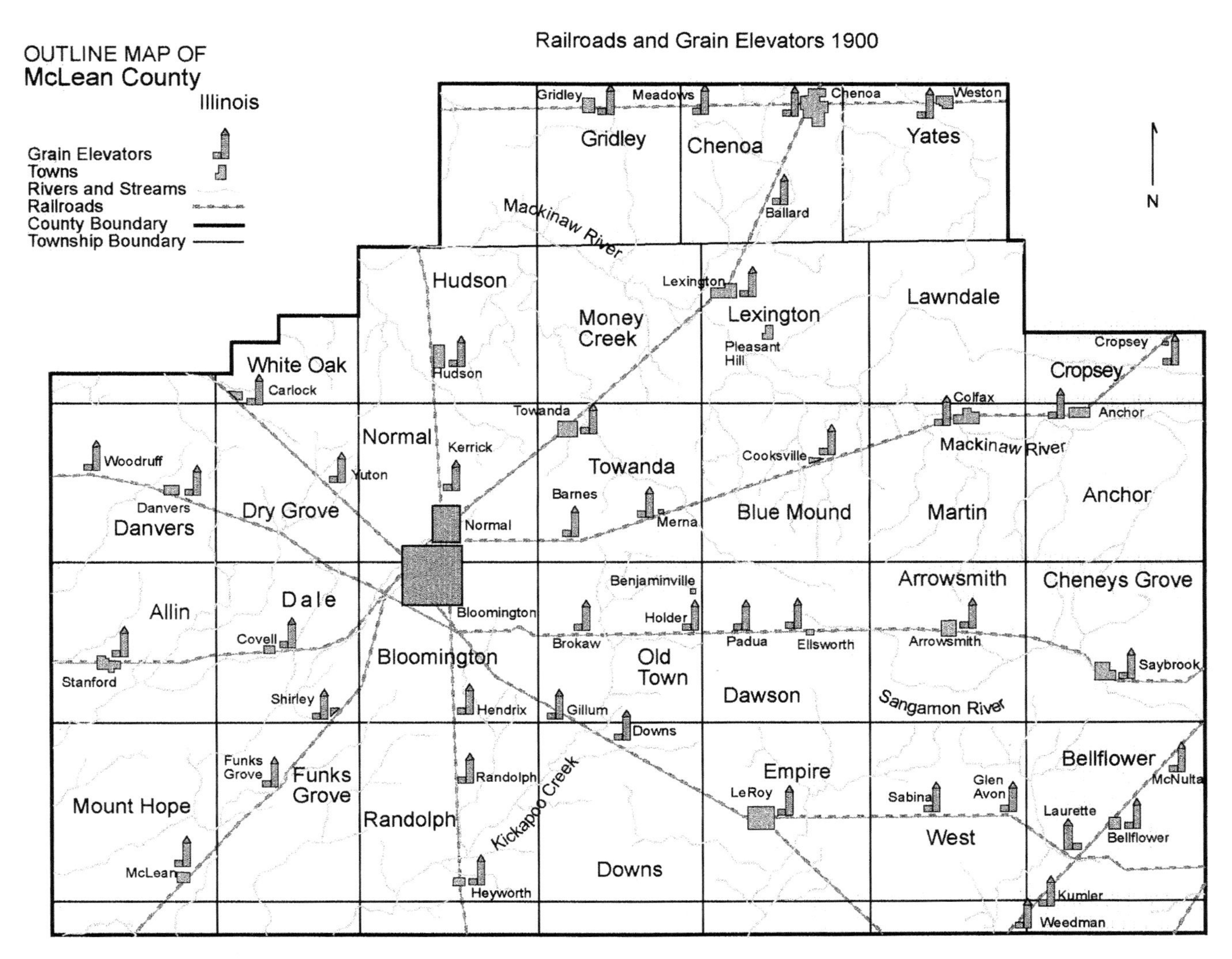

4:7. Railroads at the end of the nineteenth century. Note how they have extended into the southern and eastern part of the county. (MCHS)

In any event, Edgar's efforts were ultimately successful. The manufacture of drain tile made him a wealthy man (Heafer n.d.; Walters 1980, 71-73).

Others quickly followed Heafer's example. In the 1870s and 1880s, tile factories proliferated. Eventually, almost every small town had one, and there were more than twenty different locations in McLean County which produced agricultural drain tile. The great drain tile boom had begun. Its early intensity was remarkable. In May of 1879, John Weedman put down 7,886 feet of drain tile on his farm and it was reported, "It will not be a great while before every farmer who can afford it will drain his land in this manner, as the expense is nothing compared to the benefits to be derived therefrom." By October of 1879, the Padua correspondent of the *Pantagraph* reported, "Almost every farm in the township can boast more or less underdrain tiling put down in the last year." Farmers in the vicinity of the Old Town timber were hauling away drain tile as fast as it could be produced. The rush that started in the late 1870s continued unabated into the next decade. Millions of feet were laid down, and almost all of the tile came from local plants.

The environmental impact was astonishing. Prairie chickens, a favorite game of pioneers, disappeared. Suddenly there were fewer mosquitoes. Cases of ague diminished and deaths from malaria sharply declined (Winsor 1987, 389). Soon there was a flurry of articles in the local and national press which speculated that excessive land drainage had, in fact, greatly reduced rainfall. McLean County farmers took up their pens and vigorously denied that drain tile caused drought. They were delighted with the impact of tiling, and they had the production figures to support their gratification.

At first, Heafer and his fellow tile makers produced mainly two and a half or three inch tile, but they were soon making larger sizes. Steam engines replaced horses and tile machines multiplied. In the spring of 1880, Edgar Heafer incorporated and built a large new factory at the corner of Croxton and Hannah Streets in Bloomington. Soon, local manufacturers began to use downdraft kilns, and drying sheds. The great McLean County drain tile manufacturing boom lasted until the early 1890s. After that, farmers began to switch to tile purchased from larger factories outside the county, but laying tile continued and still goes on today. Some of the new tile is clay,

4:7a. Steam engine in McLean County about 1875. (MCHS)

but much is now plastic. Sites of many of the old tile factories may still be seen. They are usually marked by small ponds and isolated stands of timber. Most tile factories were in business for less than a generation, but it would be hard to exaggerate their impact on corn farming. Today, most of the farms in the county are tiled. In some of the flatter parts of the county, farming would be impossible without drain tile.

By 1880, changes were evident everywhere in the county. Unimproved prairie had almost vanished. In that year, there were 5,466 farms in the county averaging 134 acres each. That means, in the years between 1870 and 1880, an astonishing 1,500 farms had been added to the county's total without significantly reducing average farm size. Equally impressive was the fact that most of the farms, about 63 percent, were operated by their owners. Between

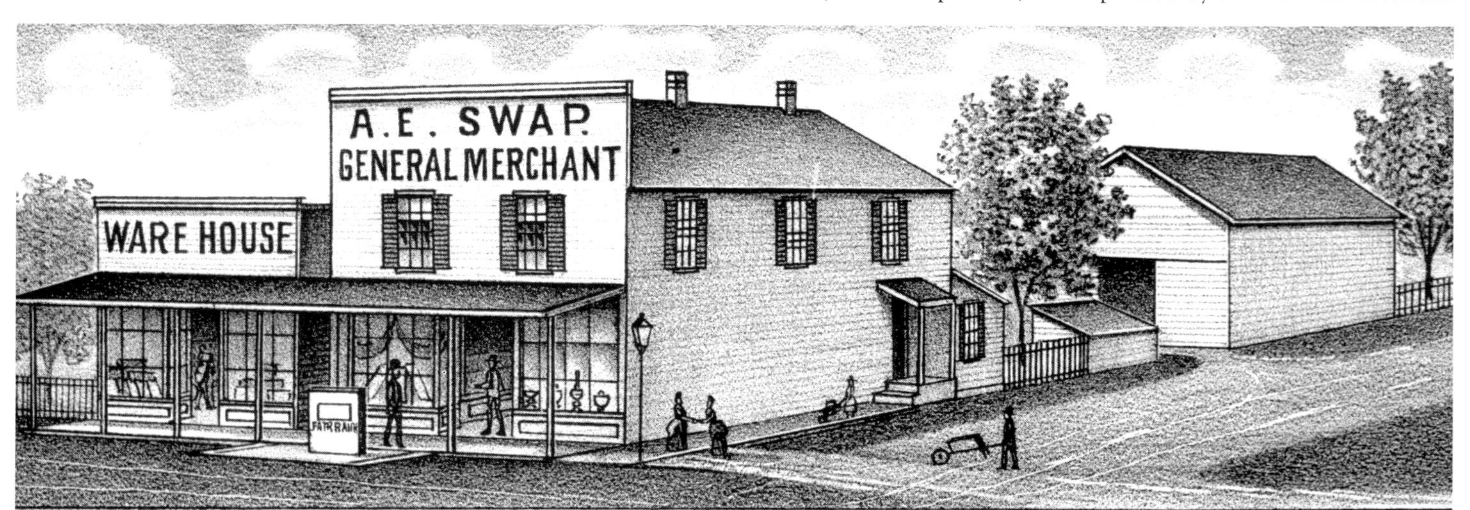

4:8. This 1887 view shows the small store of A. E. Swap in Weston. Note the Fairbanks scale in front of the store. (P & B)

4:9. Cooksville was typical of the new towns that were founded when railroads came to eastern Illinois. This 1885 view shows unpaved streets and false front stores. (MCHS)

4:10. A Cooksville grain elevator about 1910. Structures like this were designed to store and transfer grain. (MCHS)

4:11. The Merna elevator today. Once McLean County had one of the highest densities of railway tracks in the world. As these have been abandoned, elevators have either had to shut down or adjust to shipping only by truck. (W. Walters)

1860 and 1880, the number of horses and hogs had tripled. In 1880, McLean County was the wealthiest agricultural county in the state and one of the wealthiest in the nation. The 1880 census of agriculture, which actually recorded the harvest of 1879, reported that the county had produced 11,976,581 bushels of corn, three times the 1870 production. McLean County had become the top corn producing county in the state and in the nation. It retains this honor today and has held it for most of the years between 1880 and 1997. Drainage was, in part, responsible for the change, but equally important was a revolution in transportation and corn marketing.

New railroads had a particularly important impact on the eastern townships of the county. Western McLean County had numerous groves, but Grand Prairie townships in the eastern part of the county, were geographically isolated with their few stands of timber, early growth had been slow. The first burst of railroad construction in McLean County had taken place in the 1850s and had focused on Bloomington. At the end of the Civil War, the eastern townships were no better connected to world markets than they had been in pioneer days. The war had diverted resources and there was no new rail construction. The situation began to change in 1870; work on railroads began again. Between 1870 and 1895, the railroad mileage in the county more than doubled and dozens of new railway stations were established. By the end of this period, McLean County had one of the densest rural rail networks anywhere in the world and few farmers were more than four miles from the nearest tracks. In the flat eastern townships, the change was most noticeable; farms here were, and still are, larger.

The towns of Anchor, Arrowsmith, Bellflower, Cooksville, Colfax, Cropsey, Ellsworth and Holder were all established in the eastern half of the county during the post-Civil War railroad boom. In each of these towns, the pattern was the same. All handled some passengers, some freight — mostly inbound — a fair amount of livestock, and a huge volume of corn. In the 1880s, each of the towns was a shipping point for corn, and a service center for the farmers who grew the corn. The goods shipped out of Ellsworth in July of 1880 are typical. In that month, the town shipped 11 carloads of cordwood to Bloomington, 15 carloads of stock to Chicago; and 85 carloads of corn for "the eastern market" (*Pantagraph*, 29 July, 1880, 2).

One of the first functions established in each of the new "steam hatched" towns that sprang up along the rails was always a grain elevator. The railroad brought the grain elevator to central Illinois and, since the 1870s, it has remained the most important

From "Harper's Weekly," October 31, 1868

4:12. Chicago stock yards. Carloads of manure from shipped from Chicago fertilized McLean County corn. (*Harper's Weekly*)

visual element in the small town landscape. Yet, many understand little of its history or purpose. Before elevators, corn was moved to the market in sacks. In 1850, a farmer living near Bloomington would begin the process of moving his corn to market by shoveling shelled corn into hundreds of cloth bags. He would then load it, bag by bag, onto a wagon. Then the journey by wagon would begin along the mud ruts which led to Pekin or Peoria. Here the farmer would contract with a commission merchant who would arrange to ship it by steamboat to St. Louis or New Orleans. However, the sacked corn remained the property of the farmer until it was ultimately sold, and it was the farmer who assumed the risk involved in transport. This process involved an immense amount of hard physical labor. Much of the life of the miller, warehouseman and farmer was spent lugging sacks of grain from one place to another.

The invention which changed grain handling was made in Buffalo, New York, by Joseph Dart in 1842. Dart reasoned that if he could use a steam engine to drive a belt, then an endless chain of scoops could carry grain to the top of a tower, and gravity would do the rest. The problem was that Dart's much more efficient system would not work with sacked grain, and if the grain were not in sacks, how could anyone tell to which farmer it belonged? The answer was to grade corn, measure its worth by weight rather than volume, and mix the grain supply of an individual farmer with that of his neighbors (Cronon 1991, 100-132). By the 1850s, such systems were rapidly being put in place, and by the 1870s, they were nearly universal. Abuses were many, and farmers frequently complained of rigged scales and wink-of-the-eye grading, but there were also great advantages to the new way of handling corn. The logic of the grain elevator permanently altered the Cornbelt. It is a logic linked with the physics of grain and the physics of the railroad.

Even those who have spent their entire lives in central Illinois often fail to understand why grain elevators assume their

McLean County Anti-Monopolist.

Devoted to the Interests of the Farmer, the Laboring Man, the Mechanic, and the Merchant.

VOL I.--NO. 42. SAYBROOK, ILLINOIS, THURSDAY, OCTOBER 9, 1873. OLIVER C. SABIN, Sole Editor and Proprietor.

4:13. Masthead from Anti-monopolist newspaper briefly published in Saybrook. The railroads had not been in business long before many farmers came to regard them as enemies. (MCHS)

A GRANGERS' PROCESSION AND MASS MEETING.

4:14. A protest meetings of Grangers shown in *History of the Grange Movement,* a book widely circulated in the county. Things never quite reached this stage in McLean County. (MCHS)

distinctive shapes. From the standpoint of the engineer, small grains, like shelled corn, behave neither like solids nor like liquids. The grains exhibit an effect called arching which means that part of the weight of the grain is transmitted downward and part is transferred to the walls. In practice, this means that fairly thin walls can hold a substantial amount of grain, permitting early grain elevator builders to construct tall wooden structures with fairly thin plank walls. These walls were generally strengthened with bands of iron. The plank walls were then covered with some form of galvanized metal sheeting which was often referred to simply as "tin." The metal protected plank walls from weathering and helped prevent the elevator from fires started by the sparks which spewed from passing trains. The elevator was equipped with a scale for weighing grain and a steam engine which was used both to elevate the grain and to drive auxiliary machinery such as grinders and corn shellers (Carney 1995, 3-16).

Steam-powered corn shellers, which separated kernels from cob, greatly facilitated the transformation from shipping sacked corn to shipping corn in bulk. On July 29, 1880, the *Pantagraph* noted that in the newly established town of Colfax, "There are two horse power and one steam sheller at work in the town shelling corn. The Illinois Central had to send a special train down Tuesday to remove the corn." Individual farmers also purchased similar machinery. Two weeks earlier, the paper noted that Ezra Claflin had bought a steam engine to run his threshing machine and corn sheller "which will be a great relief to horse flesh." Mechanically shelled corn, which was often stored in small town elevators, quickly became the norm. Between 1855 and 1880, dozens of new grain elevators were built in McLean County.

Grain elevator location was a controversial topic closely linked to the development of many new towns. While railroads could never operate efficiently with bagged corn, they were ideally suited to haul shelled bulk corn from elevator to market. There was a limit to the frequency with which trains could stop. Railroad physics were an important consideration. Profits from railroads depended, in part, on keeping trains moving as much of the time as was possible. Long halts for loading meant loss of time and revenue. Moreover, transport by rail is efficient in inverse proportion to the time the train spends accelerating and braking. From the railroad engineer's standpoint, the ideal train would never have to slow down, speed up, or stop. From the farmer's standpoint, hauling grain is most efficient when it has to travel by wagon the shortest possible distance, so the theoretical ideal would be a station on each farm (Carney 1995, 3-8). There was a third point of view. The storekeeper cared little about transportation efficiency, but was keenly aware of the need to locate where farm families gather with

4:15. Reuben Benjamin in the courthouse's law library. His legal thinking was a major factor in the Granger Cases which changed American constitutional law. (MCHS)

time on their hands and money in their pockets. The compromise between these three positions explains the geography of post-railroad McLean County towns. The railroad companies initially decided that regular stops should be placed every six or seven miles, with seasonal halting places half way between. Each company had its own policy, but they shared in a desire to make the farmer and storekeeper pay as much as possible for the privilege of being served by their railroad.

Before construction began, the railroad would ask residents of each township to issue bonds to pay for the road. Such bonds would later be repaid by taxpayers; only a few of the early railroads, including the Illinois Central, were financed by land grants. These township bonds were often re-sold in financial centers in order to provide cash for the labor and material needed to build the roads. Few realized it at the time, but it was a system which invited abuse. As a part of this arrangement, the railroad would agree to place at least one station in a township. However, the exact location of the station was up to the railroad company. Anyone wishing the townsite on his land was expected to pay the railroad for the privilege. This payment might take the form of cash or town lots. Farmers grumbled about both the bonds and the blatant sale of station sites, but the advantages of having a nearby railroad seemed at first to outweigh the problems.

Farmers had cause for complaint and discontent was brewing among other groups. By 1880, cheap Midwestern grain was profoundly altering the course of world history. Throughout Europe, an entire aristocratic class gradually became aware that the foundations of its power were crumbling. For a thousand years, European society had been based on the assumption of food supplies that were both limited and local. Control of farmers and the food which they produced meant political control. To put it another way, in a food-short world, political power was linked to land ownership. American farmers shattered this link. By the late nineteenth century, Europeans were eating American bread and meat, freeing them to challenge timeless assumptions about political power. As European crop prices fell, income from rent declined. As farmers left the land, landlords lost their base of political power. Long established landed gentry found their incomes shrinking and their laborers leaving the land for hourly wages in the factories of Leeds, the coal mines of the Ruhr, the gunworks of Liege or immigration to America. Into the resulting political vacuum stepped democrats and demagogues. It was a fundamental political revolution which had been caused in large part by American grain but it passed almost unnoticed by McLean County farmers. They had other things to worry about.

Perceptive men were already deeply concerned about the environment. Eleven million bushels of corn might impress the statistically minded, but the cost was high. Well before the Civil War, there are references to what was called "the skinning system." The phrase referred to constant crop production with little thought to soil replenishment; today's farmer might call it soil mining. As early as the 1850s, farmers were expressing concern. In McLean County, soil depletion was first evident along the timber margins, and some of these areas have never really recovered their initial productivity. To some, the ink black prairie soils seemed endlessly rich, but others in McLean County were troubled. In 1868, Henry Funk expressed his concern. "The deterioration of our soil is a matter which begins to alarm the minds of our intelligent farmers . . . good stable manure is the chief thing we need, or all our artful and scientific means will fail at last" (*Prairie Farmer* 14 April, 1868, 218). By the last decades of the nineteenth century, farmers throughout the Midwest were keenly aware that repeated crops of corn were severely reducing yields.

It was not a problem which could be solved by hauling additional loads of manure. Extensive environmental deterioration and declining production were also the result of a very tough market for corn. Some of the discontent was psychological. "We have ceased our personal acquaintance with the miller, the plow-maker, and the other mechanics who supplied us with tools and implements, and the flour and the implement stores have taken their places. Cash has taken the place of barter and a line of middle men stand between us and the consumer" (*Prairie Farmer* 11 Jan, 1868, p.17). Some of it was generational. By the 1880s, the pioneer generation was passing, and young farmers were less aware of the exceptional bounty of the land and more keenly aware that their fathers had left them a legacy which included back-breaking labor, mounting debt, and increasing control of their lives by people who never shook their hands. Then, as now, it was these "middle men" who got most of the blame.

Price was at the root of the problem. In 1880, at the newly established elevator at Colfax, farmers were getting 30 cents a bushel for corn, the same price Richard Britt recorded in 1865. Chicago prices were a little higher. In July of 1880, they stood at 36 cents. Corn had sold for more in 1874 and 1875, but had been low in the early 1870s. Regardless of yearly fluctuations, the essential fact was that corn, in 1880, was worth not much more than it had been at the close of the Civil War. At the same time, the costs of land and labor were both increasing. In the second half of the nineteenth century, the wages of farm workers doubled (Philips 252). The gap between local prices and those in Chicago caused many farmers to question the marketing system. Their complaints were usually directed at either the large grain dealers or at the railroad. McLean County was at the center of popular protest. In Saybrook, Oliver C. Sabin took up the national cry and in 1873, published a newspaper called the *Anti-Monopolist*. The paper denounced "the commission merchants, who secure large profit, using their best

4:16. The Nelson Jones home south of Towanda was the first of several large Italianate farmhouses built in McLean County. (W. Walters)

4:17. On his farm, Towanda Meadows, William Duncan built a three story house designed to be seen from the nearby Chicago and Alton Railroad. (W. Walters)

efforts to keep farm produce as low as possible, themselves realizing handsome profits from commissions." Railroads were condemned, "all throughout the West, towns and counties are groaning under taxes to pay on bonds issued, ostensibly, for railroads but really for the benefit of contract and finance companies" (9 Oct, 1873).

In neighboring DeWitt County, there were near riots over the question of railroad bonds. In June of 1880, it looked for a time as if the governor would be forced to call out the militia to control mobs of angry farmers. Much more deft and lasting were the efforts of McLean County's great populist lawyer, Reuben M. Benjamin. Personally, Benjamin was a modest and gentle man, but he had a first rate legal mind and was unafraid of large corporations. Benjamin was a member of the group which, in 1867, redrafted the state's constitution. Written into the new document were provisions which

4:18. An 1887 view of J. J. Ham's 1873 home south of Hudson. Note the elaborate livestock barn. (P & B)

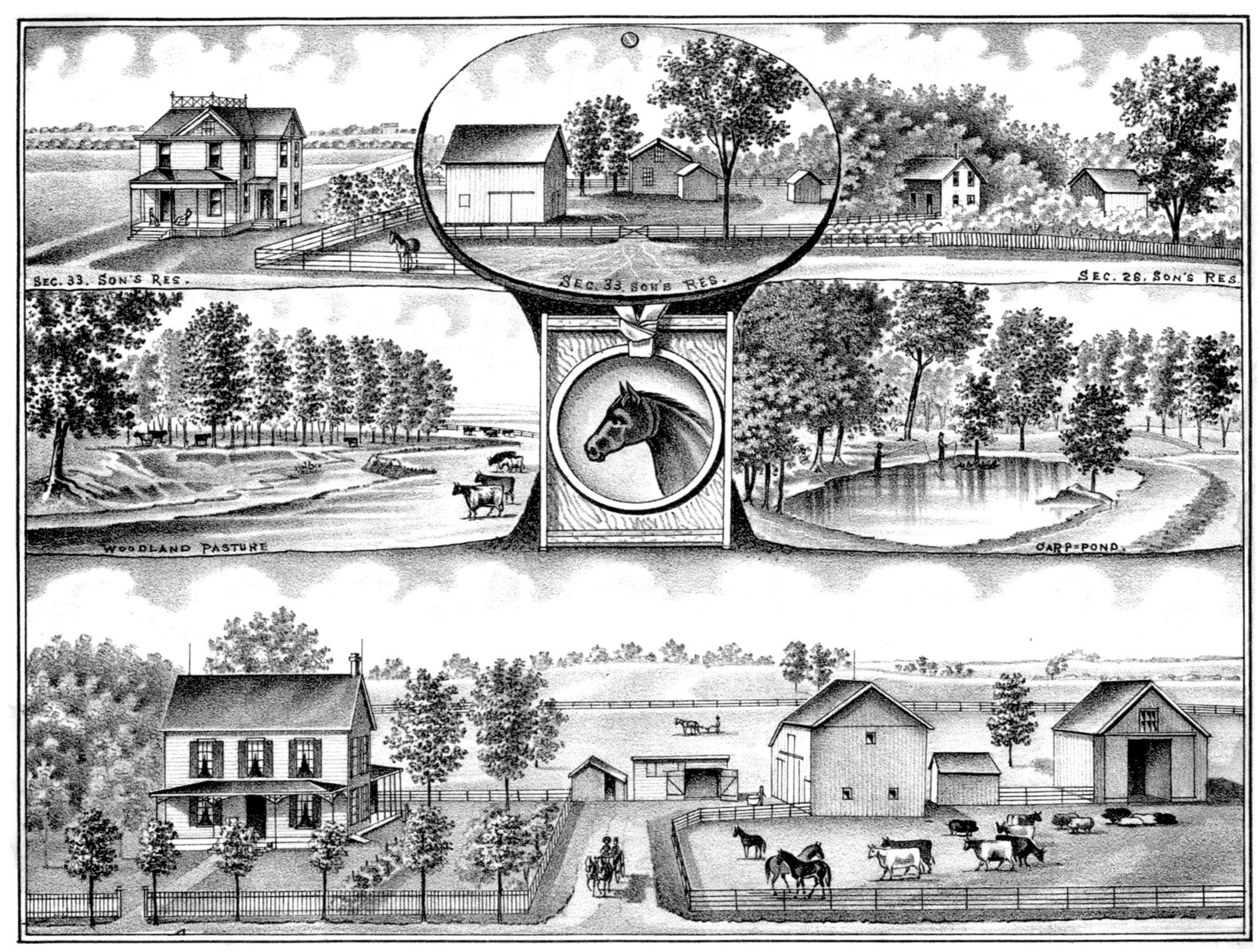

4:19. The Sholty Homestead in Dale Township, 1887. (P & B)

4:20. The P. M. Stubblefield farm in Funk's Grove Township, 1887. Note the bank barn and the wide range of fencing which is typical of the period. (P & B)

RESIDENCE OF D. L. WHITE, SEC. 21, BLOOMINGTON TOWNSHIP.

4:21. The D.L. White farmhouse is a central hall I-House. English barns, like the one shown in this 1887 view, were always more common than bank barns. (P& B)

said that the state could regulate the fees charged by railroads and warehouses. When other members of the constitutional convention suggested that such a prohibition might not be legal, Benjamin responded with a closely reasoned legal brief arguing that, while a state might grant a charter to a railroad, it did not thereby give up forever its power to exert reasonable controls over that railroad. Benjamin argued that to give up such authority was to abrogate the state's sovereignty. The subtlety of such constitutional arguments may have been lost on many farmers, but they understood his next move perfectly. Benjamin took the Chicago and Alton Railroad to court for charging more to haul lumber from Chicago to Lexington than it did to haul lumber a longer distance from Chicago to Bloomington.

Throughout the country, farmers watched as Benjamin's case worked its way up through the federal court system. Farmers in many states, especially those active in the Patrons of Husbandry, or "Grange" organizations, read and repeated Benjamin's arguments. *The People vs. The Chicago and Alton Railroad*, became one of the celebrated "Granger Cases." The Supreme Court of the United States ultimately agreed with Benjamin's logic and ruled that states could regulate railroads, and, by implication, could regulate all corporations. Grangers cheered. The case permanently altered the way in which the constitution of the United States was interpreted. McLean County became famous not just as the nation's largest producer of corn, but also as the place where the first significant battle in the war between farmer and railroad had been fought and won. Yet, even these impressive political achievements did not put much more money into the farmer's pocket.

Many farmers responded by turning their attention to livestock. If cash corn would not return satisfactory profits, why not convert corn to beef or pork? Farmers disagreed on exactly how much corn it took to carry one head of cattle through the winter, but the consensus was that a bullock could be wintered in the pasture on between one and two acres of corn. Many did just that and let the cattle take their chances outside in all but the very worst weather. Others favored stall feeding in the winter. For confined cattle, Jacob Strawn allowed half a bushel of corn per day for each head of cattle. The digestive systems of cattle are not ideally suited for extracting nutrients from corn, and much of its potential value as feed is passed out of the cattle as waste. It therefore became customary to stable two hogs with each steer or cow. The hogs rooted through cattle droppings and full value was obtained from the corn by processing it through two sets of digestive tracts (Cavenaugh 1952, p.56).

By 1890, there were 69,600 head of cattle in McLean County. Most were being fattened for shipment by rail to Chicago, but almost all farms had at least one milk cow, and there were purebred breeding herds with national reputations. For many McLean County corn farmers of this period, the final measure of success was to own a herd of registered cattle. Those who managed to obtain this status often marked their success by building large houses. For the first time, rural architecture in McLean County became more than shelter. The house became an advertisement for the success of the cattle breeder. It said to the passing farmer, "I built this house with money from purebred cattle and you too can make money by mixing the blood of your cattle with my shorthorns."

Between 1860 and 1880, the favored style for such houses is what is now called Italianate. At the time, it was simply called modern. Italianate houses feature flat roofs which extend well over the walls and are supported by fanciful carved wooden brackets. Collectively, these houses are still among the most impressive country homes in the county. In 1869, Towanda Township farmer Nelson Jones built the house he called Home Park Place shown in Figure 4:16. Jones had come to McLean County from Clark County, Ohio, in 1849 with $342, one horse, and one saddle. For a while, he farmed in partnership with his brothers and then struck out on his own. First, he raised corn and conducted general farming. Gradually, he moved into the business of breeding hogs, horses, and shorthorn cattle. The grand house he built cost $12,000. In 1873, he added a barn which cost $4,500, more than the total worth of the average farm in the county (*Portrait and Biographical* 1887, 1190-1191).

A few miles away is Towanda Meadows, which is one of the finest farm houses ever built in the state of Illinois. Figure 4:17 shows

4:22. James A. Stephens erected this fine brick farmhouse in 1877. (P&B)

this Italianate house. Located on a low hill just south of Towanda, it was originally designed to be seen and admired by travelers on the nearby Chicago and Alton Railroad. Today it can be clearly seen from Interstate 55 and still attracts a great deal of attention. Its stately design has attracted many legends, most of which are untrue. It was not a station on the Underground Railroad and there is no hidden tunnel from the house to the barn. Its owner did not keep slaves chained in the basement. The "secret room" below the floor of the second floor northeast bedroom is not a treasure chamber but a cistern for the bathtub in the main floor bedroom below. It is true that the builder's son drowned when, in rainy weather, he stepped into an abandoned well. The house was built by William R. Duncan, a unionist from Kentucky who came north to McLean County in 1863. It was probably erected in 1874-1875, shortly before Duncan's death. Like Jones, William Duncan had a national reputation as a cattle and hog breeder. He was also a public-spirited man and losing candidate for the Illinois constitutional convention. He left six children, including a daughter, Nancy, who married into the famous Dillon horse breeding family of McLean County (Sublett *et. al.* 1973, 121-122; Drury 1948, 65-66; *History of McLean* 1879, 262; *Pantagraph* 3 Oct., 1876, 3).

Figure 4:18 shows a third McLean County Italianate house as it looked in 1887. It was built by Jacob J. Ham in 1873 and still stands just south of Hudson. Like Jones and Duncan, Ham was a well known cattle breeder. The illustration is particularly interesting because it shows the multi-story livestock barn. Similar barns must have once existed at both the Duncan and Jones homes. The railroad in the distance, is supposed to represent the Illinois Central.

Other large brick farmhouses of the period include the ca.1860 Jesse Trimmer house in Money Creek Township and the 1883 Elihu Bozarth house in Allin Township. By 1890, visitors to the county were impressed by what they saw. McLean County was losing its raw frontier appearance and its landscapes were reflecting its growing agricultural wealth. This prosperity was exhibited by many middle class farmsteads as well as by the Italianate homes of the rural elite. By 1890, one could expect a substantial farmstead to include a large farmhouse, separate summer kitchen, horse or cattle barn, farm scales, hog house, chicken house, and large wooden corn crib. It would be surrounded by attractive lawns and gardens, and it would probably have a nearby orchard.

Figures 4:19 to 4:23 show some of these farmsteads. Figure 4:19 is particularly interesting, because it shows a series of buildings and farm scenes. The principal builder of the farmstead was Jacob Shotly, a native of Lancaster, Pennsylvania, who came to McLean County in 1849. The large building on the lower right is a corn crib; note that it lacks a cupola where ear corn could be loaded into the crib by movable elevator. Such elevators were just coming into popularity at the time and many of the corn cribs shown do not have this feature. In another twenty years, cupolas for elevators would be nearly universal in McLean County cribs. The large building in the lower center is the barn which was built by Jabob's son, Henry, who was a carpenter and who owned the farm in 1887. On the lower left, is the farmhouse. When Jacob first moved onto this farm, he lived in a log cabin which this house replaced about 1854. Smaller farmsteads are shown at the top of the figure. Because farmers had to pay to have the portraits or buildings drawn, small farms are not often shown. Shotley's other farmsteads give some idea of what life was like further down the economic ladder.

Figure 4:20 shows the home of Phineas M. Stubblefield in 1887. The barn is a two level or bank barn. Such barns originated in Switzerland and became common among the Pennsylvania Dutch. However, by the time of the Civil War, many farmers like Stubblefield, who had no connection with Pennsylvania, had read about these barns in agricultural journals and had copied what looked like a good idea. Once bank barns were plentiful in McLean County; now only one remains. Figure 4:21 shows a more traditional, or English barn; barns like this were extremely common in the county. Also shown are the windmill for pumping water and an extensive hedge-

4:23. The William Goodfellow farm in Dale Township as drawn in 1887. (P&B)

row. Figures 4:22 and 4:23 illustrate the efforts farmers made to plant trees around their homes and suggest the pride of successful farmers. All of these 1887 views show barns with gable, or double-pitch roofs. Within a few years, this would change and new McLean County barns would, for the most part, have gambrel, or double pitch roofs, which were designed to permit more hay storage. It's a good bet that a McLean County barn with a double pitch roof was built after 1890.

Still, the corn on which this prosperity was based remained reluctant to give up most of its secrets. It was proving to be a difficult plant to improve. Scientists examined it with microscopes, dissolved it, weighed it, burned it and carefully recorded their results. Yet, they learned little which was of value in increasing yields or resistance to disease. Part of the problem was the difficulty of increasing the density of plants in a field. Corn needed sunshine, and if one tried to plant a hill much closer than 40 inches from its neighbor, yields would go down. Even if the farmer chose only the very best kernels from the very best ears, the resulting plant often was distinctly poorer than its parent. Careful farmers worked hard at selective breeding. Perhaps the most famous of these men was Robert Reid, from neighboring Tazewell County, who developed a successful cross between Flint and Yellow Dent, which he called "Reid' Gordon Hopkins/Yellow Flint." (Hudson 1994,.56). This corn was extremely popular in McLean County and highly praised in the agricultural press. Unfortunately, the advent of Robert Reid's new corn did not make much of a statistical difference in average yields. There were hundreds of other varieties, each of which claimed special advantages; but when one looks at the overall data, it is evident that there had been no real breakthrough. In 1890, McLean County averaged 43.6 bushels of corn per acre, a yield which might have pleased, but would certainly not have surprised Richard Britt. Indeed, all of the work that scientists and corn breeders had done since 1840 had not done a great deal to increase average yields. The problem was twofold. An 1890 McLean County farmer could not thrive on 1860 prices unless he could manage to grow more corn to the acre. Moreover, thirty years of border-to-border corn had begun to attract a host of insect pests and corn diseases which made some farmers wonder if even the present yields could be maintained.

Given these conditions, it is not surprising that some farmers were beginning to leave McLean County. The change between 1880 and 1890, was small but it indicated the direction Midwestern farming would follow for the next hundred years. By 1890, there were 10 percent fewer farms than there had been in 1880. The average farm was about seven percent larger than it had been ten years before. Farmers were being squeezed off the land. The figures were not yet alarming; country folk had always known that some of their neighbors did not have enough land to compete. When old men, who had been willing to live at the subsistence level, died or moved to town, small farms were incorporated into those of more prosperous neighbors. Decline in numbers of farms was offset by an increase in the value of the remaining operations. There were now more cattle in the county than people, and over a third of these cattle had some purebred ancestry. Everywhere there was more and better machinery. In ten years, the number of horses had increased over 45 percent. Faced with such clear evidence of divine bounty, a small decline in the number of farmers must have seemed unimportant. Farmers looked forward to continued improvement. Such expectations can be dangerous, but high hopes were justified. The years to come would be the best ever for central Illinois corn farmers.

CHAPTER V

SCIENCE AND THE GROWING OF CORN
1890-1920

John H. Templin faced all of the problems of a turn-of-the-century McLean County corn farmer. Templin was a native of Kentucky who had come north to Illinois in the 1880s. In 1900, he was forty-two years old and living with his thirty-nine-year-old wife, Alice, and their daughters, Nannie 17, Menola 13, and Bessie 8. Templin rented some 400 acres near Shirley from Lafayette Funk. Templin represents two classes of farmers who were becoming increasingly common in the county, newcomers and renters. Their story has not often been told, but they were an increasingly important part of McLean County farming. By 1900, in McLean County, renters outnumbered farmers who owned their own land; 39.5 percent owned the land they farmed, 30.6 percent rented for cash, and just over 20 percent — including Templin — rented for a share of the crop. The remaining 10 percent were mostly farmers who owned some land and rented additional land. Some of the newcomers were, like Templin, from the Upland South, but others were from Germany. Almost unnoticed, German names were becoming more common in McLean County plat books. Unlike Germans in more traditional and less prosperous American communities, these German farmers quickly abandoned Old World agricultural practices. They built no quaint half-timber farmhouses and created no ethnically identifiable barn types, but they made great contributions and by 1900, they were well established.

On Wednesday, May 16, 1900, Templin had finished planting all but 50 acres of his corn. For once, the weather had been close to ideal, and one of Templin's neighbors remarked that he had never seen a better start to a corn season. The ground was moist but had proven to be easy to work. Templin's progress was typical. On that

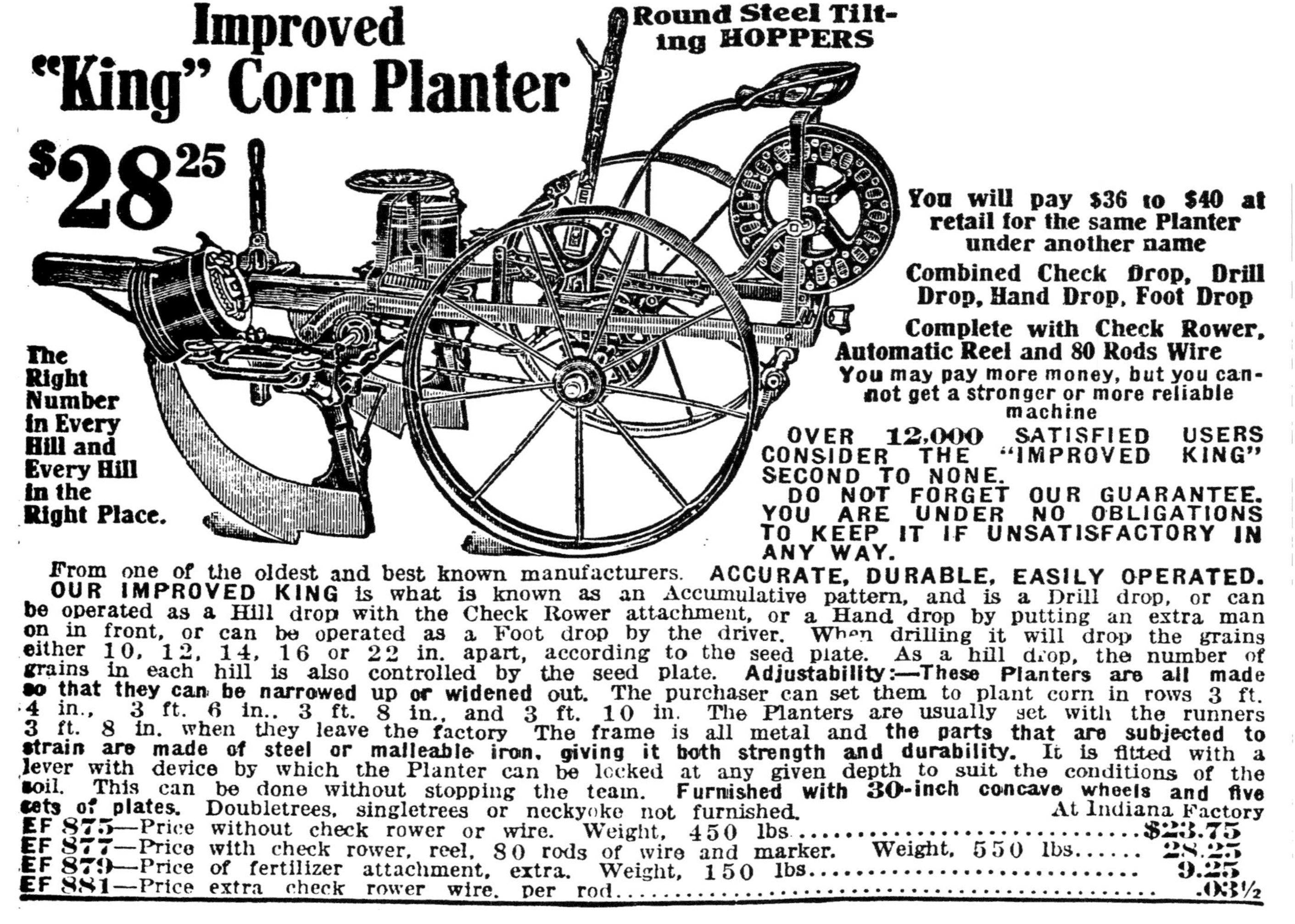

Figure 5:1. A Corn planter about 1912. Planting corn was mechanized long before corn harvesting. (MCHS)

Figure 5:1a. Plowing about 1917. (MCHS)

same day, P. S. Hulvey, who for six years had been foreman of F. M. Funk's neighboring farm, had 160 acres planted and 33 more to go. Already the corn Templin had planted three or four days before was beginning to break through the ground. By Monday, he hoped to have the entire crop planted. That is, he hoped to have that part of the crop which had been used for grain the year before, planted. There was additional corn land, which had been in sod for several years and this would take longer to plant. The existence of this sod ground is testimony to the scientific corn farmer's struggle with declining yields. Science had been enlisted in this struggle, but in the spring of 1900, the returns on this scientific investment were as yet meager.

Templin's problems began with the question of seed. Like Richard Britt in the spring of 1865, Templin was confronted with several alternatives and all of them in some respect undesirable. Following the recommendations of many farm experts Templin, and many other McLean County corn growers, refused to plant store bought seed. In 1900, there were no rules governing the labeling of corn seed. Indeed, such rules would have been impossible because there was no way of guaranteeing consistency. One agricultural writer expressed his frustration, writing, "Varietal names are of little importance." He reported, "the nomenclature of varieties is in chaos because of mixing names by seed corn dealers and the mixing of varieties by cross-pollination effected by the wind." (Hartley 1915 1-5). Seed sold under a particular name in New York might be totally different from corn seed with the same brand name offered in Illinois. The genetic make-up of the various kernels in a single sack of seed corn, even if harvested from ears grown in a single field, differed widely; most seed growers made little effort to insure consistency. Consequently, Templin selected seed corn as his forefathers had done.

Each year, he took selected ears from his own bin, hung them on a special spiked rack to dry over the winter. When planting time neared he broke off both tips of the ear. Holding the large end of the ear firmly in both hands, he pressed upward with both thumbs and loosened a double row of kernels. Care had to be taken to gently lever each kernel out of its place because what Templin would have called the "germ" of the corn, lay close to its root. Once a double row had been pried loose, the remainder of the kernels were removed by working outward. The tips and butts of ears were discarded.

Slowly, the bushels of seed began to fill. Not all of the ears Templin had saved from last fall's harvest provided him with seed that came up to his standards, and many were tossed aside. In the spring of 1900, it took Templin 400 ears to provide enough seed to plant a little over 200 acres of corn. Good seed corn, he remarked, was hard to find (*Pantagraph* 17 May 1900, 3; 19 May 1900, 8).

By 1900, a machine called the check-rower had become the standard corn planting device in McLean County. It was a development of the Brown's Corn Planter and similar machines used it at the time of the Civil War. However, with the check-rower, spacing of seeds was automatic. A farmer like Templin or Hulvey began the process of planting by driving iron stakes into the ground at each end of the prospective row. Between these, he stretched a long wire across the field. The wire was divided into segments linked by small loops, each of which held a small metal button. Button spacing varied, but three feet six inches was common. The horse drawn planter followed these wires across the field. When a knot passed through the planter, kernels of corn were deposited in each "hill." The machine could be adjusted to control the number of seeds deposited, but two or three was usual in McLean County. At the same time, a runner on one side of the check-rower marked the location for the next row. At the end of the row, the farmer would dismount, move the guide wire and repeat the process. Such devices had two great advantages. They saved labor and, because the corn plants were precisely spaced, they permitted accurate cross plowing. — at least in theory. In practice, Templin's neighbor Hulvey found that many of the "hills" were out of line, and cross plowing had to be done with great care (*Pantagraph* 21 May 1890, 4). Until the 1950s, check-rowers were common in central Illinois. Today all that remains, is an occasional strand or two of check-row wire supporting garden vegetables. Check-rowing was essential in the turn of the century Cornbelt because of increasing problems with weed control. In 1900, the only efficient way to control weeds was cultivation, and both Templin and Hulvey expected to cultivate and cross cultivate their corn at least three times after planting.

Templin and Hulvey were keenly aware of soil depletion. On what he called "old land," Templin tried to solve the problem by alternating corn with oats. He was mindful that clover, either planted by itself or "sowed," along with oats would do much to improve soil quality. Templin believed that every farmer should sow clover and that it should be plowed under in the fall. Yet Templin did not plant clover. He reasoned that "a renter can hardly afford to do so unless the land owner furnishes the seed." Templin felt that almost all renters would plant clover, if they were given seed. Unfortunately, "when a tenant pays grain rent he must be largely guided in his crops by the wishes of the owner of the land." He was blunt

Figure 5:2. Corn in the first 30 years of the 20th century in McLean County was still husked by hand. Each fall hundreds of migrant workers were drawn into the county for the harvest. (*Pantagraph*)

Figure 5:3. Husking contests often attracted thousands of spectators. This one was sponsored by Funk Brothers seed company. (Funk Heritage Trust)

about the future, and pessimistic. The land, he said, could not be kept in its present condition under the "grain system" (*Pantagraph* 9 May, 1900, 8). His views were echoed by many. A 1908 county history remarked, "the rented farms are rarely given as good care as are those cultivated by the owners." And it continued by remarking that a traveler could readily distinguish between rented and owner-operated farms by the condition of the fields and buildings (Prince & Burnham, 1908, I:737-739).

Grain rent had always been common in McLean County and would become increasingly so in the new century. Templin's arrangements were typical. In 1900, he paid his landlord, Lafayette Funk, half the grain on "new" land, 40 percent of the grain on "old" land, and half the harvested crop of hay. Templin provided his own seed corn. Funk furnished seed for the oats. Templin had been renting this farm since 1894 and was generally pleased with the arrangement. A farmer, he argued, was much better off renting than buying the land and paying a huge part of his income for interest. Of course, he admitted, the ideal situation would be for a man to have enough money in the bank to buy the land outright. Neighbors who rented for cash, paid five to seven dollars an acre for good land, three to five dollars if the soil were poorer.

Funk and other landlords were aware of declining soil productivity. They estimated that twelve years of successive corn crops cut the yield about in half and that a corn-oats rotation would limit the decline in productivity to about 25 percent. They knew only resting land would help greatly to restore yields. The damage from repeated corn crops was, in part, due to loss of soil nutrients, in part due to aggravated problems with weeds, and was also the result of increased damage by insects (Bowman and Crossley 1908, 96). Therefore, if the corn and oats rotation were discontinued, and given over to pasture for several years, the results could be impressive. In 1900, about a quarter of Templin's land was in bluegrass, what he called "meadow." For several years, such meadow would yield only hay. When eventually replanted in corn, it was called "new land." Such new land would yield well, command higher rents, and would be planted in corn for as many as five or six years in a row. After

Figure 5:4. Horse drawn corn picker. (MCHS)

this period of time, the alternating corn-oats cycle would again begin. Animal manure was still the most important fertilizer. Except for a small number of hogs, Templin raised no livestock for market. Still, his horses and hogs provided about 50 loads of manure a year. Most of the manure, he admitted, found its way onto the field closest to the barn.

Figure 5:4a. Eugene Funk. (Funk Heritage Trust)

EARLY CORN

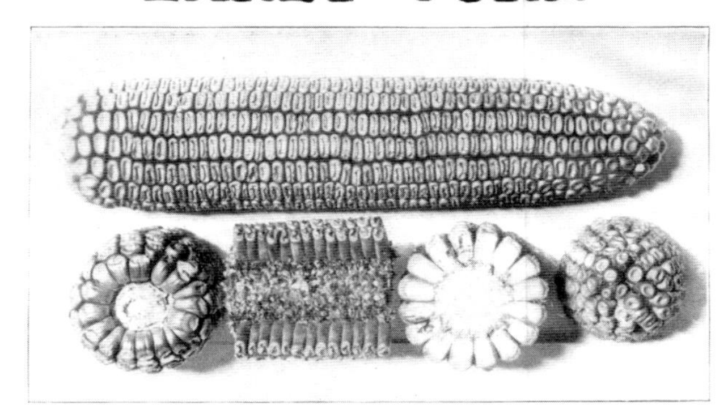

FUNK'S 90-DAY

THE EARLIEST HIGH YIELDING CORN

Funk's 90-Day Corn was originated by Mr. Eugene D. Funk in 1892 and is the only 90-day corn recognized by the Illinois Seed Corn Breeders Association as a standard variety.

The ears are good size—kernel deep—cob small. We have an early maturing corn with high yielding ability.

If you need a corn for late planting, due to hail storms, floods or droughts.

If you need an early feeding corn.

If you need a high yielding corn to hog down early.

PLANT FUNK'S NINETY-DAY—IT HAS NO EQUAL.

Funk's 90-Day fills a particular need in a wide range of climatic conditions.

The Northern farmer has in this corn an ideal silage, maturing before frost and making a large yield. For the Eastern farmer the same may be said, adding that this makes a fine feeding corn. In the South this corn has taken the place of the well known June corn making a greater yield and being a fine drouth resister.

Figure 5:5. Funks 90-day corn was one of the seed company's early products. It was bred for the northern margins of the Cornbelt where growing seasons were short. (Funk Heritage Trust)

Figure 5:6. Funk Brothers were proud of the wide range of products. (MCHS)

Templin's farm averaged between 40 and 60 bushels of corn to the acre. The previous season had been the best ever, with yields of 65 bushels to the acre. Oats averaged about 40 bushels to the acre. In May of 1900, Templin had as yet to sell his 1899 corn crop, but knew it didn't pay to be greedy. He told the story of a friend who had waited for the price of corn to go to 60 cents a bushel and wound up selling for 24 cents. Templin hoped to sell last year's crop for between 40 and 50 cents a bushel, but he had always been careful to sell before the new crop was harvested. With corn in the range of 37 to 38 cents, he would wait a little longer. What kind of income would this provide? At least a rough estimate can be made. At 40 cents a bushel, Templin and Funk would split about $4,000 for the corn crop, half that much for oats, plus a smaller amount for hay. From this amount, Templin would have to deduct all of his expenses. Clearly, a man could survive and feed his family by renting 400 acres in 1900, but it would be a long time before he would be able to afford his own land.

A farmer living a few miles east of Templin provides a somewhat different perspective on turn-of-the-century McLean County corn farming. John Karr had a 355 acre farm and specialized in livestock. When he had purchased the farm in 1877, it was badly run down. Years of repeated corn crops had driven yields on part of the land down to 25 to 30 bushels an acre. He planted oats, followed by rye, clover, and timothy. Karr hauled massive amounts of manure, 300 wagon loads a year. Other land was left in pasture and manured naturally by cattle. Even with such intensive efforts, it took the land ten to twelve years to recover. Karr's operation differed from Templin's, because he expected to make most of his money from cattle. Each year, he bought four lots of cattle and fed them for four or five months. Karr argued that even if cattle brought only the "market price," they made up for it by increasing the fertility of his corn. He summed up by saying that he couldn't keep the farm in its present condition without raising stock.

The 1899 McLean County corn harvest, which was reported in the 1900 census, had been monumental. Corn production was close to 16,000,000 bushels. Once again, McLean County led the state in corn production, but county farmers were a little chagrined when they learned that McLean, which had been first in the country in 1870, 1880, and 1890 had been edged out in 1900. Everyone agreed that the year had been exceptional. McLean County produced more corn than the combined total of 16 states. Harvest hands were hard to find and one witness to the crop wrote, "It is quite possible that the county will never again repeat such an enormous harvest of both corn and oats" (Prince and Burnham 1908, 740).

Figure 5:7. Funks early seed corn was packaged and shipped in crates on the ear. (MCHS)

Figure 5:8. Inbreds were to prove the keys to the eventual development of hybrid corn. (Funk Heritage Trust)

Figure 5:9. Picnics at Funk Farms were extensively used to promote improved seeds. (Funk Heritage Trust)

In many ways, the 1899 harvest was deceptive. It was more an indicator of excellent weather than of improved farming. Even in this year, average yields were still only about 49 bushels per acre. Machinery was much more common than fifty years earlier, but it was confined to certain phases of corn production. Planters were better, cultivators and plows somewhat improved, manure spreaders were becoming common, and thanks to a McLean County firm, portable elevators for moving ear corn from wagon to bin were numerous. However, one key part of the operation was essentially unchanged. Harvesting was done much as it had been for a hundred years. In 1900, McLean County corn husking was still done by hand. Each fall there was an invasion of men and boys with burned skin and iron hands who were drawn from every part of the United States. They joined thousands from McLean County to take part in an ancient ritual.

As their fathers and grandfathers had done, they walked beside horse drawn wagons. With metal hooks in hand, they manually ripped the corn from its husk and tossed it into the bed of the wagon. The wagons were fitted with narrow iron rimmed wheels. In wet weather, these would sink deeply into the cornfield, or when the surface of the soil had frozen, they would suddenly break through the hard crust and sink into the mire below. When this happened, an extra team of horses would be attached and the corn filled wagon yanked forward. It was commonly calculated that each harvester could gather about 70 bushels of corn a day. For this work he received two and a half cents a bushel and board. Those who supplied their own food and lodging got three cents a bushel. That men would cross the nation to earn two dollars a day for hard sun-up to sun-down labor is a stark reminder of the value of human labor at the start of the present century. Numerous corn picking machines had been invented, but as one text book on corn growing stated, "Most of them have proved impractical or wasteful." (Bowman and Crosley 1908, 203). Four to six horses were required to drag early corn pickers through fields and the machines missed much of the corn. Problems were particularly severe if the oily black soil of the county was wet.

In part, the failure of early corn-pickers was mechanical, but part of the fault lay with the corn plants. Cornstalks, especially when dry, are susceptible to wind damage. Any strong wind, such as the storms which swirled through Shirley late in the 1900 growing season, would topple many stalks. Ears on such downed stalks would usually rot before they could be harvested. In the fall of 1900, Hulvey was forced to plow under a good number of fallen ears as well as a substantial number which had rotted without falling. This was called "dry rot" by farmers and blamed on excessive heat. It would be years before corn breeders turned their attention from growing plumper ears to the equally important task of developing corn plants with stronger stalks. In 1900, McLean County was generally fortunate. Harvest weather in October was excellent. Corn and oats both did well. Templin's yield is unknown, but his neighbors were getting from 40 to 80 bushels of corn to the acre. Fifty to 60 bushels were typical.

Figure 5:10. Testing seed germination at Funk Brothers Seed Company. (Funk Heritage Trust)

In one respect, 1900 was one of the last years of the old system for growing corn. Farmers still grew much of their own seed corn. The only commonly applied fertilizer was animal manure. Weed problems were increasing, and the only answer was to cultivate and cultivate again. Seed maggots, chinch bugs, wireworms and corn root worms and many other insects thrived in fields where corn followed corn. No doubt Hulvey, Karr and Templin would all have considered themselves "scientific farmers." They were certainly aware of the abundant literature on better farming. Unfortunately, in 1900, science did not as yet have answers they needed. However, work was being done on the question of reliable seed. A key figure in this transformation was the son of Templin's landlord, young Eugene Funk.

By any standards, the Funks were an extraordinary family. Earlier, it was noted that Isaac Funk had arrived in McLean County in 1824 and had immediately begun raising corn and livestock, buying land, and fathering an extraordinarily gifted group of children. Isaac had eight sons and one daughter. Eventually, his family would accumulate 25,000 acres of the finest agricultural land in the world. Their holdings, centered around Funks Grove in the southwestern part of the county, included substantial amounts of lovely rolling woodland and huge tracts of black prairie. The Funks were an extraordinarily close knit family and those who moved to town retained an interest in the soil. Of all of the sons, Lafayette was the most devoted to the improvement of agriculture and the only one to live full time on his farm. By the late nineteenth century, he was one of the best known farmers and livestock breeders in the country. Lafayette advised governors, was president of the State Board of Agriculture, helped plan the agricultural exhibits at the 1893 World's Fair, organized farmer's institutes, and traveled the nation preaching the gospel of improved agriculture. Yet he remained a mud-on-the-boots cattleman and corn farmer.

Lafayette's eldest son, Eugene Duncan Funk, was born on September 22, 1867 on his father's farm near Shirley. He grew up in the country, surrounded by a myriad of Funk kinfolk and steeped in an atmosphere alive with schemes for agricultural reform. At age seventeen Eugene briefly attended the Wyman Institute in Alton, Illinois, but this was a temporary expedient. For the sons and grandsons of successful Midwestern pioneers, elite schools in the eastern United States had a powerful attraction. Eugene was the first of several cousins who were sent east to finish their education. He

Figure 5:11. Funk warehouse in the early part of the century. (Funk Heritage Trust)

went first to Andover, outside Boston, and then to Yale. The common emphasis at both schools was still overwhelmingly on the classics, but Eugene's taste ran to science and athletics. At Yale college, he was enrolled in the Sheffeld Scientific School. Rather than graduate, he decided to tour Europe with two cousins. The young men visited all of the requisite sights. They kissed the Blarney Stone, gawked at the towers of Warwick Castle, and admired the Rhine. Eugene had a fine time, but farms and farming were always in his thoughts. His remarks on visiting a German farm were typical." The plows, the funny things, I cannot describe them. There is a sort of cart runs along in front — pulled by horses or cows — on wheels and behind this runs a beam, sort of this fashion. . . . The funniest looking plow, I ever saw — worse than the forked stick we saw in New Mexico" (Quoted in Cavanaugh 1959, 64).

Eugene made one visit which was not on the usual Grand Tour. He went to the Vilmorin Seed Farm near Paris, where he saw agricultural seeds produced on a scale, and developed with a degree of scientific precision, undreamed of in Illinois. At that time Vilmorin Farms was run by the third generation of its founding family. They had developed the modern sugar beet and doing so had created the worldwide beet sugar industry. Eugene was deeply impressed with both the scientific techniques used and the distinctive family based management system

Back in McLean County, Eugene Funk farmed, soon married, and began a lifetime of agricultural experimentation. His first effort, modest by later standards, was aimed at developing a variety of corn which matured rapidly and could be used for replanting after early spring frosts or washouts. The result was Funk's 90-Day Corn. The process of breeding the corn made Eugene more aware of the problems of creating new corn varieties and began his exposure to the equally troubling problems of marketing improved corn. Farmers had learned to be skeptical of seed companies. Efforts like 90-Day Corn were impressive, but Eugene knew that a truly successful seed company would have to be based on long term investment and close working ties with the scientific community. He set out to put together an Illinois version of Vilmorin.

The result was Funk Brothers Seed Company. The company reflected both Eugene's ideal of scientific seed production and the distinctive Funk family social structure. Eugene began by gathering all of the information he could on corn genetics and by accumulating samples of pedigree corn from all over the country. Next followed an extended period of discussion between Eugene and Lafayette

Funk, along with Funk cousins, uncles, neighbors, and anyone else who was willing to talk about the seed corn business. In 1901, he was ready to begin operations. Thirteen cousins gathered for the first meeting of stockholders. Eugene was to be president, L. H. Kerrick — a son of Lafayette's sister Sarah — was vice president, J. Dwight Funk was secretary, and Frank Funk was treasurer. All agreed to pool their many resources and to work cooperatively in the development of better agricultural seeds. Given the large number of people involved and the tendency of families to fragment over questions of money, the organizational structure must have looked to outsiders like a formula for corporate chaos. In fact, it worked quite well and provided an exceptional degree of administrative and financial flexibility. The latter was particularly important, because it would be a long time before the seed company began to show consistent profits (Cavanaugh 1959, 87-88; 101-119).

One of Eugene's first actions was to hire, as company manager, the man he considered to be the leading corn breeding expert in the country, Perry G. Holden. Holden was a graduate of Michigan State University and a student of William James Beal, the first person in the country to successfully cross varieties of corn for the purpose of increasing corn yield. Holden was with the company only three years, but he helped to establish a legacy of exacting record-keeping and careful attention to the basics of genetics and chemical analysis of corn. To begin the breeding process, Holden began by purchasing Reid's Yellow Dent. Meanwhile, the Funks set out to visit each of the 25 leading corn growers in the Midwest and returned with samples of their best grain. The corn was stored in J. Dwight Funk's barn, where it was carefully weighed, measured, and its exact chemical content determined (Cavenaugh 1959, 87-89). The seed company then confronted the problem of accidental pollination.

Toward the end of July, when the male blossom or tassel matures, it scatters thousands of grains of pollen into the wind. At the same time, the female blossom which becomes the ear, has begun to push out between corn stalk and leaves. Each stalk normally has two ears, but anywhere from one to seven are possible. Each time a grain of pollen encounters one of the filaments on the ear, fertilization occurs and the core of a single grain of corn is formed. Anywhere from 300 to 1,000 grains form on a particular ear. Some of this pollen, of course, falls on ears of the same plant and the corn is self-fertilized. Some of the remainder spreads onto nearby plants and may drift a considerable distance. If no pollen is present, kernels will not form. In 1901, these processes were well known, but difficult to control.

The Funks began their breeding by selecting isolated plots on the various Funk farms. Here they were fortunate in being able to draw on substantial amounts of land and plant on fields which were widely separated. Isolation was essential in order to contain the random spread of pollen. Each of these isolated breeding plots contained five acres and consisted of from 80 to 100 rows of corn. Any rows which did not come up to Funk's exacting standards were detasseled, in effect, castrated. Yields were carefully measured. The best were selected for planting next year's crop. Often one select ear would provide the seed for an entire row of next year's production.

Figure 5:12. Corn Dryers at Funk Brothers Seed Company. (Funk Heritage Trust)

The next step was to select new parents in order to produce the desired qualities. In the early years of Funk Seeds, size, shape, and weight were the most important qualities, but considerable attention was paid to chemical properties. Seeds were extensively tested for germination (Cavanaugh 1959, 88-91). It should be emphasized, that at this time, Funks was not producing true hybrid corn; only carefully controlled varieties of existing corn. Yields were being improved by producing corn with predictable properties and by eliminating undesirable parents. It would be several decades before the hybrid revolution and its associated rapid increase in yield would come to the Cornbelt. Funks Seed Corn was then dried, husked, and moved to seed houses from which it was gathered into warehouses for shipment. In addition to corn, Funk Brothers acted as dealers for other kinds of seeds.

In their numerous publications, Funk Brothers used science to sell corn. In her excellent book on the development of hybrid corn, Fitzgerald noted "In seed catalogues, public addresses, and interviews, the company's commitment to the systematic, scientific investigation of corn production — in contrast, incidentally, to its competitors

Figure 5:13. Gasoline tractors attracted wide interest before WWI. (MCHS)

seeming lack of such an approach — was a reassuring and compelling theme" (Fitzgerald 1990, 135). Still, excellent and historically important science was done on the Funk acres. By 1913, Funks had introduced a three way varietal cross which they called a tribred. However, in common with other experimenters, they had not yet made commercial use of the crossing of inbreeds, which was eventually to prove the key to true hybrid corn. What they had done was to establish the most important private corn breeding establishment in the world. Equally important, they shouldered a large share of the burden of convincing farmers of the value of high quality, scientifically bred, seed corn.

The years between 1900 and 1918 witnessed a transformation of the rural landscape of McLean County. No farmer ever likes to admit prosperity. Talk over coffee is usually of problems, and it would be pushing one's luck to describe a great year as being more than "fairly good." Yet, corn prices in these years were remarkably good, even when adjusted for the modest inflation of the period. In terms of buying power, corn was worth half again as much in the period between 1900 and 1910 as it had been a decade earlier. Things were even better between 1910 and 1918 (Hart 1986, 55). The years of the "Great War" were the best McLean County farmers had yet experienced. Before the United States entered the war, both the Allies and the Central Powers attempted to starve their enemies into submission. Food became an important weapon and corn prices soared. In the spring of 1917, Americans joined the fighting, but corn prices kept going up. Unlike other agricultural crops, corn remained unregulated. By October of 1917, the price reached $1.17 a bushel. Even though costs were a little higher, the bottom line was in favor of the farmer. In the fall of 1917, huskers were being paid five cents a bushel plus board. That year, D. O. Thompson, the county's agricultural agent, estimated that between 750 and 1,000 migrant laborers were needed to bring in the corn crop. The McLean County Better Farming Association placed advertisements in newspapers in southern Illinois, southern Indiana and Kentucky, informing workers of husking opportunities. Some farmers complained about paying such high wages, but most were keenly aware they were far better off than in the days when huskers were given three cents a bushel for 38 cent corn.

Between 1900 and 1910, the value of McLean County farms more than doubled. In the next decade, it increased by another 80 percent. The 16 million bushel 1899 corn crop, which many had regarded as unbeatable, was exceeded by the 23 million bushel crop reported in the 1910 census, but this yield was a truly exceptional. The 12 million bushel 1919 crop was more typical. Oats also did well in both these years. To be sure, each decade saw a steady drop in the number of farmers and a steady increase in the proportion of tenant farms. In 1920, over half the county's farmers rented their land, but there were still over 1,700 owner-operated farms, and most of these were free of mortgage debt.

This 1900-1918 prosperity is strikingly reflected in the landscapes of McLean County. Anyone who travels the township roads of McLean County today has to be impressed with the number of large and substantial buildings dating from the first two decades of the twentieth century. One cannot understand the development

Figure 5:14. Among the new materials to become popular in the early twentieth century was concrete block. Farmers purchased machines like this to make their own blocks. (MCHS)

Figure 5:15. Turn of the century prosperity saw corn farmers building many new houses like this large concrete block structure south of Carlock. (W.Walters)

Figure 5:16. Barns too were constructed of concrete block. (W.Walters)

Figure 5:17. Concrete fenceposts were cast around metal frames and lasted longer than wooden posts. (W. Walters)

of corn farming in McLean County without realizing the importance of these years of good prices. It was the golden age of the Illinois Cornbelt. Everywhere, the countryside was alive with new ideas and, for once, farmers had the cash to put these ideas into practice. The visible reminders of these good times were new buildings.

Rural housing was transformed. Hundreds of new homes were erected, most of them large, square structures with central chimneys, houses which have since come to be called "American Foursquares" or "Corn Belt Cubes." Today such houses are one of the defining landmarks of the Illinois landscape. The square shape of these houses was a result of central heating; their smaller windows, evidence of electric light. They differ strikingly from nineteenth century houses because of higher foundations and more substantial construction materials. Far more than their predecessors, they were factory made houses with windows, doors, porch columns and cabinets hauled ready-made to the construction site. Farm families were proud of their new dwellings. The *Pantagraph* commented on this revolution by reporting that the farmers were putting into their houses "every convenience and feature." They were "multiplying throughout the county" and rural people could boast, "we are building houses just as good as any in town" (13 Oct. 1917, 3).

There was much more. New barns were being built in great numbers, most of them designed to house cattle. Hundreds of new corn cribs were erected. In the fall of 1917, the lumber company in

Figure 5:18. The years of high corn prices between 1900 and 1920 saw the construction of rural houses like this one. They are often called Cornbelt Cubes or American Foursquares. (MCHS)

Figure 5:19. Hollow building tile was another improvement and particularly popular for the new oval corn cribs. This one was built by the Springer family about 1915. (W. Walters)

the tiny town of Cropsey reported that it had sold lumber for seven new cribs and there were plans to supply the material for four more. In form, these were like their predecessors, with two slatted cribs for ear corn on either side of a central passage and storage for threshed oats or wheat above the drive-in passage. There were two important differences. These were large structures, 30 by 36 feet. More important, almost all were equipped with both interior and exterior elevators. On the farm, elevators for moving ears of corn into and out of cribs, were one of the most important and rapidly adopted innovations of the period. "Scooping," one observer remarked, "was practically a thing of the past" (*Pantagraph*, 13 Oct., 1917, 17).

There were also forests of new silos. Traditionally, corn had been used in one of three ways: stored as ears, shelled for grain, or left in the field for animals to forage. Silage provided a new option. Entire ears, husk and all, were chopped and blown into what were, in effect, giant storage cans. Even a modest silo might hold 80 wagons of corn. In the late 1870s, Dr. Manly Miles, at the University of Illinois, published a book recommending that farmers make more use of silage, but the idea was slow to catch on. Some silos began to appear on Illinois farms in the 1880s, but it was not until the twentieth century that they became common features in McLean County. A major reason for this was that Cornbelt farmers spent large amounts of their new income on cattle. Hogs paid the rent, but cattle offered prestige. Silos provided winter feed for these newly purchased cattle. They preserved feed that would otherwise be lost and kept cattle fed when summer pastures were burned dry. In other areas, silos were often associated with dairy farms. There was a rapid increase in dairy cattle at this time, but many McLean County silos were built on general purpose grain farms to feed beef cattle. Scores of silos were built in 1917 and 1918.

This was a period of experimentation with new materials. The most important of these materials was concrete or, to be more exact, Portland cement. Because McLean County lacks building stone, it has always been a county of wooden buildings. Locally, burned brick had always been used for the foundations, but it was poorly suited for this task and would quickly flake and crumble, especially at the groundline. The alternative was to bring in yellowish dolomite from quarries near Joliet or whiter limestone from quarries to the north. Both were temporary expedients, expensive, and not a great deal more durable than local brick. Insurance companies demanded brick for retail buildings, but the farmer lived in a world of wood. Therefore, it was hardly surprising that, when quality concrete became available around 1900, farmers in McLean County became major consumers of the new material. They built retaining walls, barnyard floors, chicken coops, hog houses, tool sheds, milk sheds, windmill foundations, well houses, cattle troughs, cisterns, boundary markers, stock tanks, culverts, bridges, storm cellars, silos, corn cribs, and even an occasional barn out of the new material.

Figure 5:20. Tile was also popular for silos. Marvin Hougham stands next to the silo built by his father about 1920. (W. Walters)

Figure 5:21. The tile Price Jones Barn near Towanda was built in 1914 to house feeder cattle. (W. Walters)

Much of the concrete used was formed by the farmers themselves. Walter Weinheimer recalled the time his mother's cousin came out to the farm to help with the construction of a hog house. "He stayed here, of course we had a gravel pit down here. Dad bought a block machine and he made all the blocks for that. . . . Yes, handmade blocks. You put the gravel in there and the pallets and packed it in by hand pulled levers . . . turned them over, took them out, then you set them out. Set them to cure under a tree . . . He stayed one whole year. That's all he did was make a few blocks every day. He was getting up in years pretty well" (McFIP).

Concrete fence posts were extremely popular. In 1917, there were about five billion fence posts on American farms, and each post had an average life of about five years. Repairing and replacing fences consumed huge amounts of farm labor. Wood and metal fence post forms were commercially available, but many farmers made their own. With a small cement mixer or even a trough, the farmer would mix local gravel and local sand with cement. Generally, the posts were reinforced with wire rods or wire. End and gate posts were heavier, with horizontal footings, and hinge foundations often imbedded into the concrete. A good farmer and his helper could make 25 posts an hour. When the job was done, they could use their mixing tools to pour cement blocks (Ekblaw 1917, 146-166). Throughout McLean County, many handmade early twentieth century fence posts still stand and function as well as they did when first erected.

The other revolutionary building material was hollow building tile. After 1900, local drain tile factories began to lose their market to larger national firms. A few decided to salvage their operations by producing hollow glazed blocks for building. Unlike concrete which blends imperceptibly with its surroundings, a tile building provids a magnificent splash of red to the landscape. Such tile blocks might be used to build houses, hog sheds, or silos, but their most important use was in oval cribs for ear corn. In these cribs, some of the tiles were perforated to allow free passage of air to remove moisture. Like wooden cribs, they were built around internal elevators and designed to be filled with a second portable elevator. Access was through a roof cupola or, at ground level, by means of double wooden doors. William Springer of Stanford, was so impressed with these structures that he built three of them on his land, and two of these survive. Unfortunately, these curious structures were more visually impressive than functionally efficient. Most of the surviving examples have long been empty and their numbers are rapidly dwindling.

Even more spectacular were tile barns. In 1914, Price Jones constructed three such barns near Towanda. They were designed to feed

Figure 5:22. In 1916 William Naffziger built this circular tile sided barn. It is seen here about 1972. (W. Walters)

Figure 5:23. The lower walls and foundation is all that remains of this tile sided round barn north of Normal. Like the Naffziger barn it was constructed around a central silo. (W. Walters)

Figure 5:24. This building on the Leo Miller farm was constructed as a residence in the 1870s. Within twenty years it had been converted into part of a sheep barn. (W. Walters)

beef cattle, and they incorporated the latest in construction technology, using cast concrete rather than wood for many of the supporting pillars. Even more impressive were circular tile barns. Three round tile barns with central silos were constructed in McLean County. One was built by the Funk Brothers and has been gone for many years. A second was constructed by Conrad Schaefer, northeast of Normal, but only the walls remains. Most impressive of all the huge structures, was the one erected by William Naffziger in Mt. Hope Township. It was made of locally burned tile and arranged so that cattle faced inward toward the tile silo. This was the last surviving circular barn in the county, and it remained intact until its roof was ripped off by a tornado in 1996. The walls and silo still stand.

Farmers also spent their new money on less visible improvements. Corn went north to Chicago and returning cars were loaded with the daily deposits of hundreds of thousands of big city horses and the waste from stockyards. Funk Brothers used Chicago equine manure to help grow their seed corn. In September of 1910, neighborhood farmers gathered in Holder to unload 25 car loads of Chicago brown. At 40 tons to a carload, this amounted to small mountains of the fertilizer. Charles Yancy was delighted with his 525 tons and calculated that would be enough to cover 30 acres. Manuring was good farming. So were improved roads, and almost every township in the county had launched a program to construct what were then called "hard roads." It was also good farming to educate your children and, for the first time, substantial numbers of farm children were enrolling in degree programs or short courses in the state's new agricultural schools. The county now had a government agricultural agent and it was not quite so insulting to be called a "book farmer."

There was also the rage for tractors. Some McLean County farmers were beginning to acquire tractors, especially between 1915 and 1920, but they were still far from universal and the range of activities they could perform was limited. In McLean County there was a long wait between the advent of tractors and their general use in corn farming. Charles Hart and Charles Parr built the first successful gasoline engine farm tractor in 1901. County farmers were fascinated, but skeptical. The first factory-sponsored demonstration of a gasoline powered tractor in McLean County took place on the Raymond Dooley farm in 1914. Two years later, in August of 1916, there was a week long tractor extravaganza on a truly massive scale. It took place on the East Lawn Stock Farm a mile and a half east of Bloomington. One hundred exhibitors demonstrated 350 of the new machines. Five hundred acres were plowed. Over one thousand men were employed in connection with the exhibit, and total attendance was close to 100,000. It was one of the largest agricultural events yet held in the county (*Pantagraph* 21 Aug., 1916, 7).

Figure 5:25. The replacement house on the Leo Miller farm reflects the great agricultural prosperity of McLean County early in the present century. (W. Walters)

Long distance telephone and telegraph lines were installed for the tractor show. There were special street car connections and exhibition trains from nearby cities. Ten Boy Scouts were on hand to carry messages. The Women's Exchange Club served thousands of lunches. Farmers were impressed. J. A. Everson remarked, "Cash corn is now selling at the highest figure ever known, 86 cents a bushel. The average crop is not likely to be increased very much. So we turn to our operation costs for a possible reduction in the expense of farming." Horses, Everson argued, were the farmer's most expensive tool, and oats to feed them consumed huge amounts of potential corn ground. Viewed in that light, a tractor didn't seem like such a large investment. Farmers swarmed over Allis-Chalmers, Big Bulls, Deeres, Corn Belt Tractors, Daunchs, Happy Farmers, Hart-Parrs, Fords, Illinois Tractors, Pol Parrots, Simplexes, Standard Detroits, Sweeneys, Waites and a large number of Averys made in Peoria. There were six sizes of Avery tractors, all run on kerosene, and they were "backed by an established company with a large factory and many branch houses which insures your getting a well built machine and permanent and prompt service after you get it" (*Pantagraph* 22 Aug., 1916, 9). McLean County entered the tractor age with boundless enthusiasm.

As the war continued, the price of corn continued to move upward. Each time German defeat was predicted, corn futures would briefly quiver, but when the rumors proved false, corn would again resume its climb. In October of 1918, Number 2 Yellow Corn, sold for cash in Chicago, was bringing farmers $1.52 a bushel. As the price of corn went up, there was a corresponding increase in the value of McLean County farmland and in the rent which landowners charged their new tenants. Labor, however, was still a problem. In 1918, huskers were hard to find, and many farmers were forced to pay hands seven cents a bushel. Some even suggested using women huskers, but the idea was not well received. Somehow, the corn was eventually cribbed, and it proved to be another excellent crop. In November, when the armistice was signed, everyone knew that the price of corn would fall, but farmers generally expected a modest decline. Perhaps prices would stabilize at something like the average for the immediate pre-war decade. The farmers had reason to be confident. After all, they were the best educated, best housed, best supported, and best equipped farmers in the history of the world, and they were farming the best corn growing land in the world. Fortified in their new houses, proud of their new livestock, they looked out over newly filled new bins, awaited the return of their sons and hired hands, and faced the future with confidence. They were in for a devastating surprise.

Chapter VI

CELEBRATION OF CORN

By the twentieth century, corn had become more than an agricultural commodity. It had become a focus of community celebration. Corn had always been a featured item at county fairs, but gradually it also became a centerpiece for non-agricultural entertainment. At first, corn shows were small town events with close links to both local farmers and corn growing associations. The activities at these early corn fests were much like those at county and small town fairs. They were rural events, organized by local families, featuring home-grown decorations, governed by modest budgets. Later corn shows changed. They, moved to larger cities, cost more to put on, and the emphasis shifted from promoting corn to promoting the places which sponsored the shows. These festivals constitute one of the most colorful chapters in the story of McLean County corn growing.

The Shirley Corn and Horse Show, held on October 13 and 14, 1911, was typical of early corn festivals. To house the show, families from Shirley and nearby Funks Grove erected a huge tent capable of holding 1,000 people on open ground between the Christian Church and the Shirley School. Everyone was pleased with the outcome of the 1911 show and it was decided to repeat the event in 1912 and to make it a three day celebration. In 1912, two poles decorated with corn stalks flanked the entrance of the tent. Between the poles was the sign proclaiming, "Shirley Corn Show." Officially its title was the Shirley Corn and Horse Show, but horses got second billing. Wise in the ways of central Illinois weather, the locals made certain that the tent was heated. Many community families took part including the Funks. DeLoss Funk was vice president of the event and Eugene Funk was a member of the Exhibits and Premiums Committee. There were also many other local participants including John Templin, the tenant farmer mentioned in the previous chapter. Special trains ran between Bloomington and Shirley. Oratory addressed domestic science, hard roads, and corn. Musical entertainment included the Illinois State Normal University Glee Club. Except for the horses, all products had to be grown or crafted in McLean County.

The presence of Leigh Maxey at the 1912 show is indicative of the growing importance during these years of corn judging. Maxey, who was from Curran, Illinois, was a nationally known corn judge. He had been a judge in many national corn shows. On the first

Figure 6:1. The 1912 Shirley Corn and Horse Show. (Ruth Carpenter)

Figure 6:2. Inside the tent at the Shirley Corn and Horse Show. (Ruth Carpenter)

day of the fair, Maxey conducted classes in corn judging which both men and boys who had entered corn in the show were expected to attend. The corn they submitted was divided into classes. Classes "A" and "B" were for ten ears of white and yellow for exhibitors over 18 years old. Classes "C" and "D" were for "boys and girls." There were also awards for the best bushel of corn in each class — a bushel being reckoned to include 50 ears — and sweepstakes for combined ranking in several classes. There were also other categories. These classes are interesting because they indicate the qualities progressive corn growers were trying to promote. For example, Class "J" was for the best hill of corn including stalks, ears, and roots. Class "L" was for the stalk producing the greatest amount of shelled corn. Class "O" was for the best ten ears of 90-day corn. Ears of corn were judged using 100 point scorecards; points were given for qualities like uniformity, shape, market condition (that is, was it ripe?), tips, butts, kernel uniformity, kernel shape, length, circumference, and proportion of grain to cob.

The Shirley exposition was more than just a corn show. There were the usual additional fair entries for items like doughnuts, apple pie, plum jelly, canned gooseberries, late potatoes, hominy, ketchup, pillow cases, candy, and dresser scarves. The Shirley Corn and Horse Show was all a great deal of fun, but marked by a deeper purpose. It was no accident that the rise of corn festivals coincided with a growing national awareness that there were serious problems in the countryside centering on the lure of the city. By 1912, rural people were deeply aware that bright lights and better prospects were beginning to attract many of their most ambitious young people.

To rural people, it was not just a question of lost labor, it was a matter of morality. All over the country, politicians were deeply concerned about the flight from country to city. In 1907, when Eugene D. Funk became president of the National Corn Growers Association, he saw fit, in his address at the National Corn Exposition in Chicago, to quote extensively from recent remarks by president Theodore Roosevelt. Roosevelt had urged farmers to stay on the land, learn to grow more grain and at the same time to retain the fertility of their soil. In spite of such rhetoric, the flight to the cities became even more pronounced. Rural festivals were also becoming increasingly city-dominated events and these events, in turn, were being influenced by forces from outside the community. These trends are evident in the history of the Bloomington Corn Festivals.

Long before corn festivals, the citizens of McLean County's largest town had celebrated their connection with corn. In the summer of 1900, the center of the town was destroyed by fire. In the following months, it was quickly rebuilt with newer, more fireproof, and often larger buildings. The tallest of the buildings erected in the year after the fire was the Cornbelt Bank. To decorate the columns next to the main entrance of the new building, architect George Miller decided not to use any of the classical Greek motifs which were then popular. Instead, stone carvers were brought in to sculpt the tops of the red Lake Superior sandstone columns into ears of corn. The capitals remain today as a symbol of the county's most important crop.

The first of two huge corn festivals was held in Bloomington in the fall of 1915. In theory, the Bloomington Corn fests were

Figure 6:3. The first of two Bloomington Cornfests was held in 1915. This is the Corn Palace. (MCHS)

Figure 6:4. The 1916 Corn Palace. This year the festival was held earlier to avoid bad weather. (MCHS)

Figure 6:5. The interior of the 1916 Corn Palace showing the giant corn dollar. (MCHS)

harvest festivals. As such, they could only be held after the corn was husked. As a practical matter, since most of the decorations were to be made from corn, they could not realistically take place until late fall. This timing was to prove most unfortunate. The 1915 Corn Fest was held from November 1 to November 6 and was to coincide with Bloomington Prosperity Week when many stores featured sales. Co-sponsors of the 1915 event were the Commercial Club of Bloomington and the McLean County Better Farming Association. It was to be held at the Bloomington Coliseum, an exhibit hall located at the corner of Front and Roosevelt. The hub of the festivities was to be a strange, hermaphrodite building known as the Corn Palace.

The idea of a Corn Palace was neither new nor original to McLean County. Many Midwestern towns had built such structures, but the sponsors of the Bloomington event felt that as the biggest corn growing county in the nation, they certainly ought to have the biggest and best Corn Palace. So, the Bloomington Corn Palace was advertised as "The most prestigious corn palace ever built in Central Illinois." The palace was a temporary structure, created by rebuilding the exterior of the Coliseum. A timber false front was added to the front of the coliseum which was covered with decorative swirls and spirals formed by thousands of ears of corn and supplemented by other farm products. Between 800 and 1,000 bushels of corn covered interior and exterior surfaces. Additional decorations included three wagon loads of baled alfalfa, two wagon loads of loose alfalfa, five wagon loads of corn stalks, three wagon loads of pumpkins, and one wagon load of Sudan grass. The outside of the Palace was brilliantly lit by a 1,000 watt Pearless nitrogen lamp and 11 smaller 750 watt lamps which were hooded to cast their light on the corn-covered walls of the building. One observer recorded, "The show looked great in the afternoon, but the brilliance of artificial lights increased the passing effects, and the consensus opinion was that the display was even more beautiful in the evening than during the day" (*Bloomington Bulletin*, 1 Nov. 1915, 8). Perhaps it looked good in the glaring artificial light, but it was an undistinguished period in American architecture, and the best that can be said for the design of the Corn Palace is that it was not a great deal uglier than the surviving Corn Palace at Mitchell, South Dakota.

Inside, the Corn Palace was certainly spectacular. Upon entering visitors were confronted by a titanic ear of corn, 30 feet long and seven and a half feet around. The framework of the huge ear was made of wood, but its surface was composed of hundreds of ears of corn. A protruding cob and a great deal of imitation corn silk completed the immense ear. The platform on which it rested was divided into compartments and in these were displayed the ten ear entries of hundreds of "boy farmers." Further inside, the visitor could view domestic exhibits, a map of McLean County made of corn, and the Rotary Club's 12 by 16 foot corn-clad symbol surrounded by a gold colored-frame.

Figure 6:6. Many Bloomington merchants featured advertisements linked to Cornfest exhibits. (*Pantagraph*)

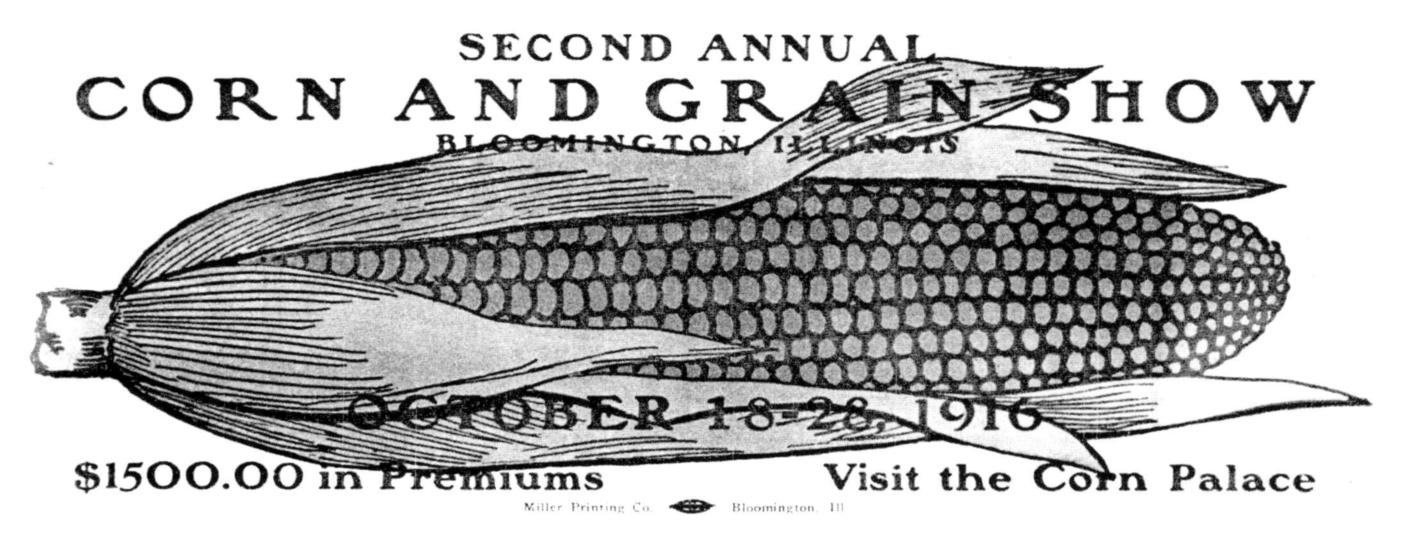

Figure 6:7. Blotter from the 1916 festival. (MCHS)

Even in 1915, mass media altered the events it reported. This is clearly shown by what happened on November 3, 1915. On this day, the junior corn was to be judged, but it was also the day on which filmmakers from Pathe arrived to make a newsreel about the Palace. The newsreel crews immediately decided that not enough was going on. Tranquil views of a corn-coated exhibition hall would not play well with film audiences. They asked the organizing committee if something couldn't be done to provide a little more action. After all, they were offering nationwide exposure for the Corn Palace. The committee thought things over and then went to the Superintendent

Figure 6:8. Plowing contests were popular attractions. This one was held in 1914. Arthur Moore remarked that there was nothing like the sound of an engine to attract farmers. (MCHS)

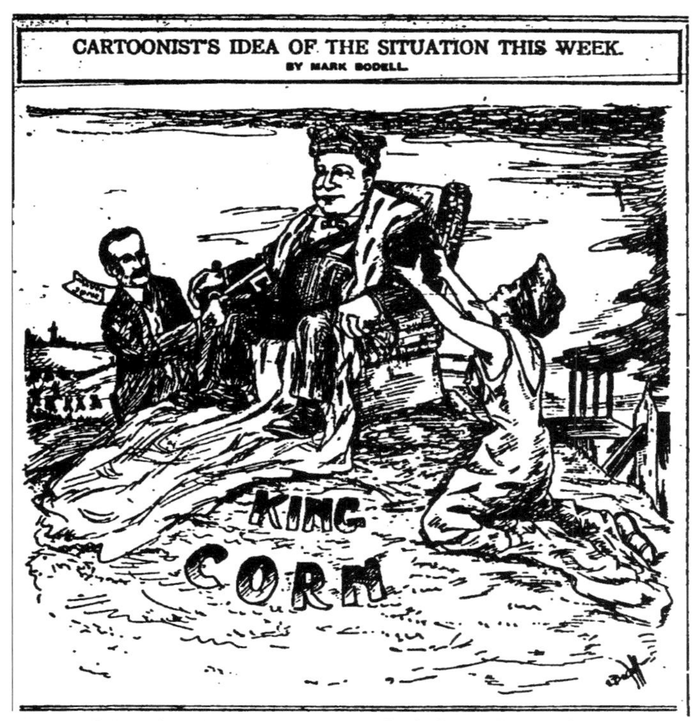

Figure 6:9. King Corn from the 1916 Corn Show. (MCHS)

of Schools. He was persuaded to dismiss students, fourth grade and above, an hour early so that they could "lend animation" to movies of the Corn Palace. For its part, the committee agreed that after three o'clock that afternoon the usual 15 cent children's admission fee would be suspended. The Bloomington Band was assigned the role of pied piper and led thousands of children downtown to the Palace. At the appointed hour, students arrived and stormed the exhibit hall. From his perch on a wagon bed, the delighted cameraman cranked away as the students swarmed into the building. Confusion was such that corn judging was postponed until the next day. The Bloomington Corn Palace had its moment of fame, and kids had an unforgettable day.

Still, nothing could overcome the Illinois weather. Rain kept the crowds small. Overall paid admissions were only 26,000, impressive numbers to be sure, but considerably less than the committee had hoped for. Bloomington merchants in particular were pleased, so it was decided that although the 1915 event had lost money it would be worthwhile to repeat it in 1916.

The 1916 festival was even larger and lasted longer. It was advertised as the Second Annual Corn and Grain Show and $1500 in premiums would be awarded for the best entries. It was hoped that the festivities could attract over 50,000 people. The 1915 corn show had been billed as the biggest thing in Central Illinois; to top that, the 1916 event was going to be "the biggest thing of its kind in the Midwest" (*Bloomington Bulletin* 15 October 1916, 2). Again the focus was to be on a corn palace built around the Coliseum. The 1916 Palace was designed by Bloomington's most prominent architect, A. L. Pillsbury. Its facade was a pattern formed by thousands of yellow ears of corn trimmed with dried Sudan grass, and gentlemen were strongly admonished not to smoke their pipes, cigars or "cigarets" near the Sudan grass. Over the door, huge letters proclaimed, "Corn Is King." Show promoters were proud of the fact that every-

Figure 6:10. Giant Ear from the 1915 festival. (MCHS)

Figure 6:11. Pfister float in the 1947 Corn Bow Parade. (*Pantagraph*)

thing had been designed and erected by local people. To avoid the previous year's problems with bad weather, the 1916 festival began two weeks earlier and would run from October 18 to 28.

All of the corn to decorate the palace was to be donated by McLean County farmers. By October, organizers were worried that because of the earlier date they would not have enough ears to do the job. Bloomington papers published a plea for the hundreds of additional bushels of white, yellow, and red ears needed to finish the palace. Farmers responded and by October 15, everything was "Loaded to go off in apple pie order" (*Bloomington Bulletin* 15 Oct., 2).

Bloomington was famous as one of the places which had pioneered the post card. Accordingly, Joe Burt printed 20,000 color post cards of the Corn Palace which were offered for sale. The German American Bank promised to donate 20 dollars in gold for the heaviest ear and to display that ear in their bank lobby. Exhibits came from all over the Midwest. C. T. Williams of Farmer City exhibited an ear of corn with 21 rows of kernels, a genetic freak because an ear of corn was always supposed to have an even number of rows. W. H. Brigham crafted a huge dollar coin, 30 inches in diameter, made entirely from kernels of corn and other small grains. The corn dollar survives to this day in possession of the McLean County Historical Society and is perhaps the only major remnant of either Corn Palace. In 1916, there were more exhibits, including over 1,000 corn entries. As before, there would be vaudeville; however, the acts were carefully screened by a committee to assure they were appropriate for family viewing. Visitors could watch Dyer and Peters, thrill to the Diving Nymphs, listen to Lew Fitzgibbon on the xylophone, and gape at Havemann's animals. A major 1916 addition was to be dancing, which would start at 9:30 each evening. Organizers promised that, except on Republican and Democratic Day, there would be no political speeches during the vaudeville and dancing. The corn festivals were not rowdy events. They were family affairs, carefully controlled, with a strong educational component.

Inside the Palace were hundreds of exhibits. There was serious corn judging at both festivals, and there was a great deal of serious corn science. In 1916, Lyle Johnson of the Farm Experiment Station contrasted the very different producing qualities of two ears which were visually identical. One of the first speakers was Theodore Kemp of the Rotary Club's Good Roads Committee, who echoed the growing national cry for investment in hard surface rural roads. The University of Illinois sent a huge model farm, so large it filled an entire railroad car.

Figure 6:12. Sommer Bros. Seed Float from the 1947 Corn Bowl Parade. (*Pantagraph*)

Children played a central role in the corn festivals. Every rural school in the county sent some kind of exhibit. In 1916, there was a grand automobile parade with colorful banners and 90 cars containing 874 parents and children, representing each of the country schools. There were also numerous contests. For example, the boys-husking contest was held on Friday, October 27. A huge number of unhusked ears were piled on the Corn Palace stage. Teams of two boys, competing for the best time, were then required to husk 75 ears. Judges stood by to count the husked ears and to make sure that each one was fully stripped. Charlie Killion won the 1916 event with a time of four minutes and 20 seconds.

Unfortunately, the weather was again dismal. About seven on the opening evening of the 1916 show, a heavy rain began to fall and it continued through the night. Muddy roads delayed promised speakers. Only a third of the expected five or six thousand first day visitors filed into the Corn Palace. What was politely called "deplorable weather" continued for the first week of the festival. It was the "worst October weather ever experienced" and the roads leading to the city were almost impassable much of the time. Equally disturbing was the fact that unfavorable weather throughout the growing season led to generally poor quality agricultural products and McLean County was not able to put its best foot forward (*Pantagraph* 28 Oct. 1916, 7).

Speakers at the 1916 Corn Festival reflected changing ideas on many subjects. Among these was the proper role of farm women. McLean County women had always worked in the corn fields, although certain jobs, like planting and husking, were traditionally reserved for men. For a pioneer family like the Baldridges in 1850, it was acceptable to admit that women joined in the process of growing field crops. In fact, pioneer farm families generally took great pride in the agricultural roles of women. By 1916, this attitude had changed. As farm families sought middle class acceptance, they were forced to recognize that the middle class of the early twentieth century was reluctant to admit families with wives who performed certain kinds of manual labor. For farm wives, this kind of work included what was called "field work" to distinguish it from farm tasks like feeding chickens which took place closer to the home. Almost all of the tasks directly associated with the growing of corn fell into the category of field work, which is not to suggest that women did not continue to perform this kind of labor in McLean County; they did. Field work, however, was not encouraged by the reform minded and middle-class oriented agricultural press, and in truth, women were often reluctant to do such work. A talk given at the Corn Festival on October 29, by G. I. Christie, a Purdue University professor, illustrated a growing vision of a farm wife who did not labor outside the house and garden.

Christie discussed child rearing on the farm. He outlined to Corn Festival audiences how a farmer should help his sons learn the various farm tasks. He then turned to the subject of farm girls and reminded his audience that girls on the farm should not be forgotten. The farmer should raise his daughters to make a good housewives and to learn "something" of a farm. What was it a farm girl should know? Well, said Professor Christie, "When Jack raises a carload of cattle and tops the market she should know enough to know that he has achieved something. When he wins the grand national sweepstakes at the fat stock show, she should be able to clasp him by the hand and know what it means. He is a first class business man. I don't mean she should milk, feed the stock and plant the grain but she wants to know something of farm life" (*Bloomington Bulletin*, 29 Oct 1916, 4).

Final accounting showed that the 1916 Corn Festival lost between two and three thousand dollars. Especially disappointing was the small number of rural people who attended. Organizers promised a new and better show in 1917, but war intervened and the tradition was not continued. Some wanted to keep the Corn Palace building as a memorial. Unfortunately, pigeons found the exposed ears of corn to be a major attraction, and the Palace had to be dismantled. The pigeons have remained.

After World War I, no more corn palaces were built in McLean County, but there was an abundance of agricultural celebration. Husking contests drew huge crowds, and the winner often husked well over a ton of corn. Farm machinery was a part of these early expositions, and pulling contests with tractors and horses were popular throughout the twenties and thirties. Politicians were plentiful at such events, but so were farm machinery salesmen. Fortunately, county farmers have always trusted machinery more than they have trusted congressmen. In 1945, Arthur Moore wrote, "A great man might be soaring on the wings of eloquence . . . But let a motor cough somewhere on the grounds and farmers would stream away from the platform like hounds after a mechanical rabbit. If Moses himself had arrived on the platform to the roar of thunder and the flutter of a thousand angels' wings, and with a revised and popular edition

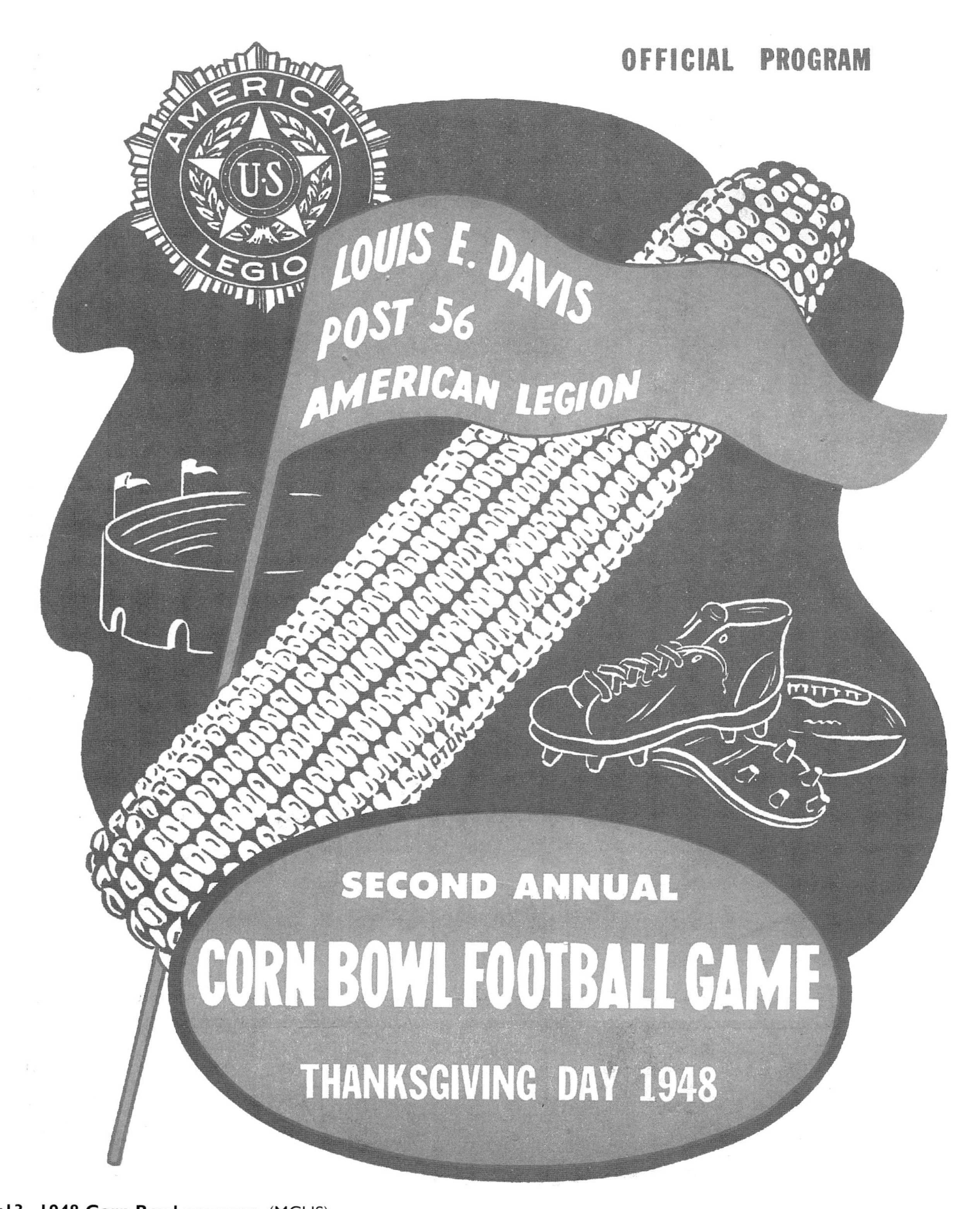

Figure 6:13. 1948 Corn Bowl program. (MCHS)

of the Ten Commandments, I think farmers would still have followed the sound of a gasoline engine" (Moore 1945, 134).

The failure of the Corn Palaces did not end urban efforts to celebrate corn. By the late 1930s, local businessmen decided to try another corn-centered event. In October,1939, Bloomington hosted what they called a Corn Belt Exposition. There was to be a sort of corn palace, although this one would be more modest than its predecessors. The theme of recreation dominated, and humor was much more in evidence than in the earlier events. Visitors were encouraged to, "Take a vacation. Travel to the Kingdom of Corn. See the King of Corn in Person. Witness the Coronation of King Corn and the Queen of the Corn Belt. Play in the Street of the King's Jesters and the Street of the Little Kernels. Visit the Corn Palace and learn more about the products which have created and which

Figure 6:14. Feasting on sweet corn at the Corn Bowl celebration. (MCHS)

maintain this land of wealth and happiness. Participate in the celebration of a bounteous harvest. Let Joy Prevail. Long Live the King." The King's treasurer promised to provide visitors with King Corn Currency which would make them King for the day and which could be risked at gaming tables and devices. There were Ambassadors of the Seed Growers who promised to tell the story of hybrid corn, and there were Earls of Industry who explained corn processing (Corn Belt Exposition 1938).

McLean County used the Exposition to brag about its advantages. On the cover of the exposition program, surrounded by a border of plump ears of corn, these advantages are listed. The county was "In the heart of the corn belt." Its production of all cereal crops was second in the nation. Retail sales and per capita income were above the national average. The percentage of homes owned and families with automobiles was well above state, sectional and national averages. Racial attitudes were unreformed and unhidden. One advantage was that the "native white population" was more than 91 percent. This was ironic indeed because of the great role that foreign-born people had played in the county's corn growing and in face of the fact that in the late nineteenth century, African Americans who tried to farm in the county had been driven off the land. The pamphlet went on to proudly proclaim, "A day's vacation in the realm of King Corn will afford you a delightful relaxation, a wealth of information and a world of satisfaction that you live in a land of plenty."

In 1939, King Corn was Harold W. Ennis, of Minier. His queen was YWCA cashier Irene Anderson, who was described as a "chic blonde." The royal couple were crowned with appropriate ceremony and seated in thrones which had backs painted to resemble huge ears of corn. Charles W. Kirkpatrick read the coronation decree making them rulers of the Corn Belt. This interesting decree invokes a number of places and corn related events. In part, the king and queen were installed "on behalf of every ear of corn raised in the 31 townships of McLean County, the largest county in the state per square mile. On behalf further of every ear of corn raised in the other 101 counties of the state of Illinois; on behalf of every ear of corn raised in the other 47 states and provinces of the United States of America . . . And in proud remembrance of that lamented great artist Alfred Montgomery, a Bloomington son, who immortalized corn in oil paintings; and gratefully recalling the revered Civil War governor of Illinois, the Hon. Richard J. Oglesby, who perpetuated forever in the words of a tribute to corn, delivered in September, 1894, 45 years ago." On the stage with the newly-crowned king and queen were the King's Cabinet, including farm advisers and Farm Bureau presidents from 14 counties. Boy Scouts acted as a guard of honor. The coronation was followed by fireworks and a ball (*Pantagraph*, 21 October 1939).

By the 1940s, corn palaces gave way to football games. The first annual Corn Bowl was played on Thanksgiving day of 1947. From Naperville, Illinois, came North Central College, two-time champions of the College Conference of Illinois. Their opponents were the 1947 champions of the Illinois Inter-Collegiate Athletic Conference, Southern Illinois. The seven-mile-long pre-game parade included 93 floats and was viewed by over 30,000 people. It was said to be the largest ever held in the city. Memorial Stadium at Illinois Wesleyan University was decked in corn stalks. The North Central team proved no match

Figure 6:15. 1947 Corn Bowl Parade. *(Pantagraph)*

for the Maroons of Carbondale and went home with a 21-0 victory. After the game, coach Abe Martin praised the defensive play of Bunker Jones, Charles Heinz, and a young man with the name of John Corn. As usual at corn fests, the weather was dreadful. Freezing temperatures, icy streets, and snow greeted the visitors.

The second Annual Corn Bowl was played on Thanksgiving Day of 1948. This time, the home town Illinois Wesleyan Titans, coached by the school's former standout, Bob "Big Bluestem" Morrow, went up against Eastern Illinois College, from Charleston, coached by Manard "Pat" O'Brien. It was billed as "the only big Thanksgiving struggle in this part of the state" (Corn Bowl program 1948). Promotion of corn was again part of the festivities, and football fans were reminded that corn was still king and that Bloomington was the capital of the biggest corn producing county in the United States. Eastern was heavily favored, but the Titans thrilled the crowd with a last minute 6-0 victory. In keeping with the times, the Official Program was filled with patriotic fervor. The American Legion, which helped sponsor the event — reminded audiences that the legion "takes the lead in disseminating knowledge of and promoting respect for the national flag; it does important work for the boys of America through Boy Scout activities and junior athletics; it combats revolutionary radicalism and all movements which have for their aim the downfall of America."

An essential part of McLean County corn festivals was the recitation of Richard Oglesby's tribute to corn. It is an odd piece of literature and it has an odd history. There was a time when hundreds of school children were proud to have committed the former governor's words to memory. There was also a time, long ago, when teachers were wise enough to know that memorizing could be a marvelous tool for instilling confidence in normally shy students. They realized that there was an age when learning the capitals of the states, or the dates of the Presidents, is an attractive and exciting game. Indeed, there were young men and women who, without special encouragement from the teacher, committed "Horatio at the Bridge" or the Gettysburg Address to memory for no better reasons than savoring the marvelous rhythm of the words or for simple pleasure of proving to playground comrades that thay had mastered something difficult. Oglesby's "Royal Corn" belongs to that bygone era of recitation.

The governor's story of how the words came to be written is not easy to believe. However, since he has never been proven wrong, one must accept his version of how it happened. Oglesby explained that it was September of 1884 and he was at a club's Harvest Home Festival in Chicago. He was seated at the south end of the speaker's dias and deeply engaged in conversation with Conan Doyle, author of the White Company and the Sherlock Holmes stories. Each speaker was to deliver an address on the subject, "What I know about farming." When the toastmaster called Oglesby's name, the former Illinois Governor arose, realized he had nothing prepared, and looked around the room for inspiration. His eyes rested on the magnificent stalks of corn which decorated the walls. He began, without further thought, to speak slowly. Even more improbable is the fact that the speech was not written down until four years later when a member of the club, Volney W. Foster, copied it out from memory. The following lines reproduce the first part of Oglesby's oration; for the full effect, one must imagine that he or she is standing in front of a

Figure 6:16. 1947 Corn Bowl Parade. (*Pantagraph*)

Corn Fest audience, and the words must be spoken out loud in clear ringing tones rather than simply read.

"The corn, the corn, the corn, that in its first beginning and its growth has furnished aptest illustration of the tragic announcement of the chiefest hope of man. If he die he shall surely live again. Planted in the friendly but somber bosom of mother earth it dies. Yea, it dies the second death, surrendering up each trace of form and earthly shape until the outward tide is stopped by the re-acting vital germ which breaking all the bonds and cerements of its sad decline, comes bounding, laughing into life and light the fittest of all the symbols that make certain promise of the fate of man. And so it died and then it lived again. And so my people died by some unknown, uncertain and unfriendly fate, I found myself making my first journey into life from conditions as lowly as those surrounding that awaking, dying, living infant germ. Aye, the corn, the Royal corn, within whose yellow heart there is of health and strength for all the nations. The corn triumphant, that with the aid of man hath made victorious procession across the tufted plain and laid foundation for the social existence that is and is to be This glorious plant transmuted by the alchemy of God sustains the warrior in battle, the poet in song and strengthens everywhere the thousand arms that work the purposes of life. Oh that I had the voice of song or skill to translate into tones the harmonies, the symphonies and orations that roll across my soul when standing sometimes by day and sometimes by night upon the borders of this verdant sea, I note a world of promise, and then before one-half the year is gone I view its full fruition and see its heap-ed gold await the need of man. Majestic, fruitful, wondrous plant. Thou greatest among the manifestations of wisdom and love of God, that may be seen in the fields or upon the hillsides or in the valleys" (Oglesby 1894).

It was not only with palaces, football games, and poetic words that people celebrated corn. There was also at least one man who made his living by painting corn. The idea is not as strange as it may at first seem. For generations, wealthy farmers had hired well known artists to immortalize their horses, cattle, and dead deer; why not paint corn? Indeed corn with its striking green and yellow contrasts, its natural play of light and shadow between kernels, and its contrast of textures is in many ways an ideal subject for the artist. The nation's foremost corn painter, Alfred Montgomery, had strong connections with McLean County. Montgomery was born in Lawndale, a few miles southeast of McLean County in 1857; he died in Los Angeles in 1922. But in between he spent a good deal of time and painted a substantial number of pictures in and around Bloomington.

Like many painters, Montgomery led a vagabond life and, unlike some, he thoroughly enjoyed it. As a youngster, Alfred peddled vegetables on the streets of Webster City, Iowa. As he matured, he decided that painting was more fun and a good deal more profitable. He was entirely self taught. Between 1888 and 1889 he tried his hand at teaching art in the public schools. One suspects he delighted his pupils and horrified his superiors. By 1890, he was living and sleeping in a barn near Mason City, Iowa. Alfred would drift from town to farm, a brash and exceedingly confident young man, with a quick laugh and a roguish sense of humor. A man who knew him well later wrote, "I know he did not die of nervous prostration, or nervous exhaustion . . . He was no thin-skinned bird" (Bowman, MCHS archives). By 1893, Alfred had come back to central Illinois and was

Figure 6:17. After the great fire of 1900 the Cornbelt Bank was rebuilt and featured corn ears on the pilaster capitals. (W. Walters)

working full time as an artist. Unfortunately, few of Montgomery's paintings are dated, so it is difficult to establish which of his many paintings were done in McLean County. The earliest dated Montgomery painting is from 1896. From 1893 to 1897, he lived and worked at 806 E. Chestnut in Bloomington. He is also listed in Bloomington city directories for 1904 and 1905. His work was popular in McLean County. Mrs. Howard Humphries, one of Bloomington's important society hostesses, was extremely proud of her collection of Montgomery paintings and several hang in the Funk homestead (Hamilton and Hamilton 1978, 69-75).

By the turn of the century Montgomery was beginning to gain critical recognition and to win awards. His works were exhibited at the 1900 Paris Exposition. Midwestern visitors would wind their way through miles of galleries and stand in groups to admire his work. Montgomery combined the talents of an artist with those of a salesman. When the urge struck, he would wander tramp-like through the country, painting for his supper on rough boards, cigar boxes, planed lumber, tin, patched canvas or academy board. Hotel keepers in farm towns were among favorite customers, but he was never shy about approaching millionaires or bank presidents. Indeed, he was never shy about anything. No secretary or "do not disturb" sign ever stood in his way. He would trade paintings for railroad passes or whiskey, as confident of his ability to sell as of his talent as a painter. One observer thought that there was scarcely a bank in central Illinois that did not have a Montgomery painting on its wall. Seed companies were another prime market.

Why Montgomery selected corn as his subject is unclear, but most of his paintings include corn, and in many, corn is the featured subject. He once remarked that his ideal was to paint corn so realistically it would deceive a horse or a bird. Frequently he painted Reid's Yellow Dent; perhaps the dimpled texture appealed to his artistic instincts. Often the ears are in bags, barrels and battered crates and surrounded by other produce or simple farm tools. Eventually he tired of central Illinois, drifted into Oklahoma, and finally settled near Los Angeles at a ranch which he humorously christened, "Nowhere." Academic painters were not always thrilled with his choice of subjects, and no one ever accused him of having been corrupted by impressionists. He painted what he liked and what he knew. He had a grand time doing so. Perhaps the most accurate summary of his work was by an Oklahoma critic who called his paintings "folk songs" (Hamilton and Hamilton 1978, p.70).

Chapter VII

HARD YEARS 1920-1950

School children learn that the Great Depression began in 1929. Farmers know it started ten years earlier. In the years that fighting raged, it had been common knowledge that the end of the Great War would drive down the price of corn. Armistice rumors always sent corn futures tumbling. What no one understood is that when the war actually came prices would drop lower and stay down for longer than anyone believed possible. The years between 1920 and 1940 were the hardest ever experienced by McLean County farmers. Not just corn prices, but almost all farm products reached record low levels. In part, the story of corn farming during the thirty years between 1920 and 1950 is the story of these dreadful years, but — especially after 1935 — it is also the story of triumphant World War II production and the beginning of technological changes which would fundamentally alter the way in which corn was grown.

Perhaps it would be well to begin by reviewing a typical McLean County farm in the year 1920. The farm would have contained about 165 acres, almost all of which was farmed and most of which was used to grow grain crops. The farm would have been divided into from four to eight fields by hedges and fences of woven wire, barbed wire, or wood. About half of the total farmland would be planted in corn. By far the largest single source of farm income came from the sale of corn as cash grain; that is, by selling corn to an elevator or other dealer rather than using it as feed for farm stock. After harvesting, most of this corn was stored as ears on the farm in large slatted wooden cribs. When the farmer felt corn had reached its maximum price or when the farmer needed cash, the stored ears would be hauled to a local elevator, shelled, and sold. Almost all corn raised was still open-pollinated and had been grown on the farm where it would again be planted. About a quarter of the farmland was devoted to oats, which were mechanically threshed by traveling crews and then held in solid walled bins to eventually be fed to work horses, or perhaps sold as needed for a cash crop. Much of the remainder of the land would be sown as clover to be

Figure 7:1. Hedged Cornfield about 1925. (MCHS)

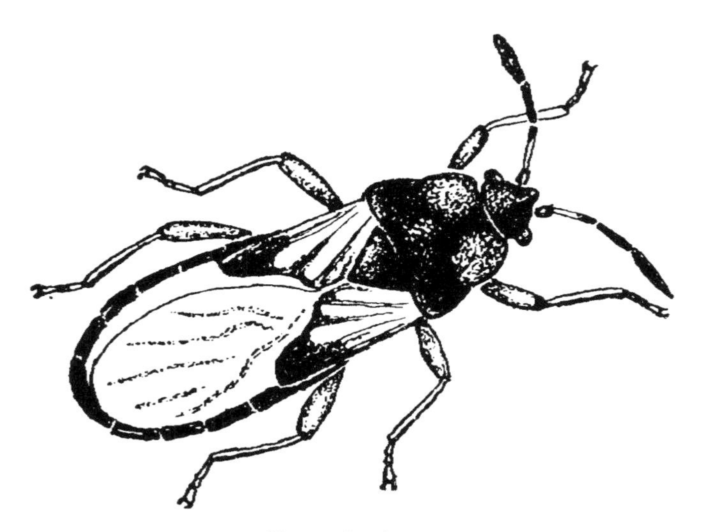

Figure 7:2. The chinch bug destroyed millions of dollars worth of corn in the 1930s. Actual size is 1/16". (MCHS)

plowed under in order to provide nitrogen for the next year's corn crop. There would also be some pasture from which hay would be cut in the summer. Crops were rotated each year. Most farmers also raised some cattle and fed a small number of hogs, which were either butchered on the farm or fattened for market. There were chickens for family use. Cattle and hogs were turned loose on harvested fields to manure and to glean crops left in the field.

The typical farm would have been operated by a tenant. However, in 1920, there were still a substantial number of owner operated farms. Travelers believed they could easily distinguish between owner operated farms and tenant operations by a quick glance at the condition of the buildings, the standard of fence repair, the state of weed control, and the quality of the flower garden. On tenant farms, the tenant would supply labor, seed corn, farm machinery and tools, and he would be expected to keep buildings and fences in good repair. The landlord would supply farmland, house and farm buildings. Long term tenancy was common; ten years was not unusual. Often there was no written contract between landlord and tenant; they would talk together about the crops to be planted and the system of crop rotation to be used. At the end of the year

Figure 7:3. Barriers erected to confine chinch bugs. (MCHS)

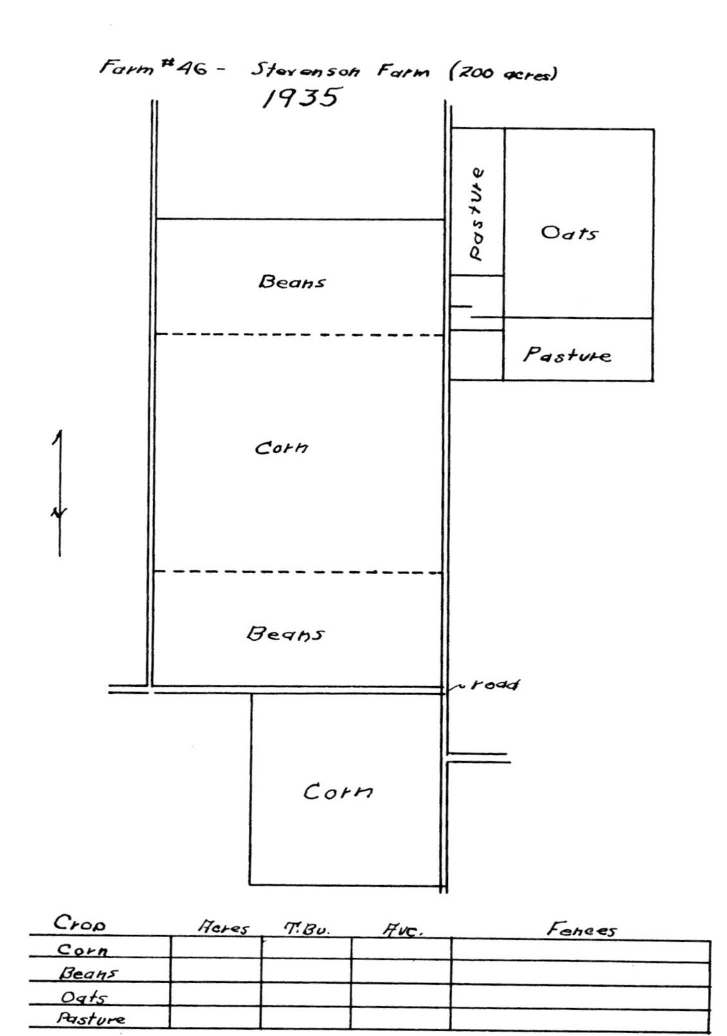

Crop	Acres	T.Bu.	Ave.	Fences
Corn				
Beans				
Oats				
Pasture				

Figure 7:4. Land use on Farm 46 in the early 1930s. (MCHS)

they split the income from grain crops. Usually some cash rent was also paid, often for the use of pasture land. Corn was planted by a horse drawn check-rower in May and harvested by hand in late October or November. All farms had horses; tractors were known and widely discussed, but they were expensive and mechanically unreliable. Other than manure, and an occasional application of powdered limestone, very little fertilizer was applied. Almost no conservation measures were taken and erosion was a serious problem.

The typical farmer lived in a one or two story frame wooden house which was probably the original dwelling on the farm. The farm usually had no electric service, but telephones and automobiles were not uncommon. Clustered near the farmhouse would be a barn used for horses, livestock, and hay storage. There would have been at least one large wooden corn crib which was almost the same size as the barn, and a number of smaller cribs. There would also have been from three to a dozen smaller buildings, usually wooden and often in bad repair. Roads were often graveled and much better than they had been twenty years before, but paved farm roads were rare. The farmer and his wife were literate, fully conversant with world and national affairs as they affected farm prices, and had numerous relatives nearby. Other than the influence of occasional pamphlets from the Department of Agriculture and a small but growing involvement with the county Extension Agent, the Federal government was uninvolved with farming operations.

The fences and hedges which surrounded the fields required constant maintenance. Mowing along fences was a required regular effort. When it was too wet to work in the fields, fathers sent their sons into the hedgerows with a short-handled, sharp-bladed instrument called a hedge knife. Hedge growth was hacked off at armpit height. Since the hedge was often reinforced with wire, hedge knives would frequently strike hidden metal and need to be re-sharpened. It was work that made a young man eager to get back to hoeing corn. There were many variations to this pattern but the general picture must be kept in mind in order to understand what happened in the following years.

McLean County farmers lived in a world where everything revolved around the price of corn. Almost every financial decision depended on this price, and it was by far the most important factor influencing the yearly income of both tenants and landlords. The price of corn also determined landlord's willingness to make farm improvements, the ability of farmers to buy new equipment, and the farm family's reaction to campaigns for higher property taxes to support roads and schools. The price of corn also powerfully influenced the need to borrow, the willingness to lend, and the price of an acre of land. If one is to understand what happened to McLean

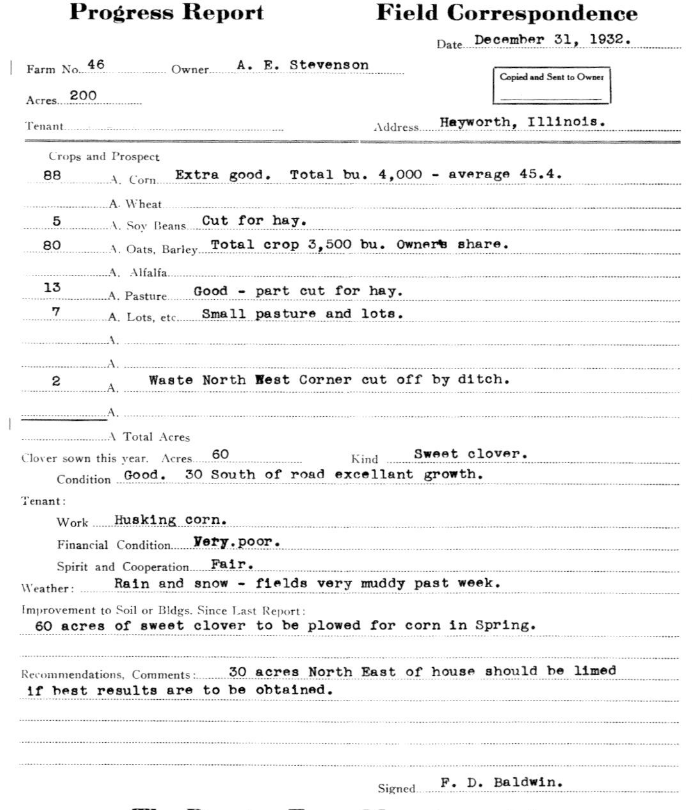

Progress Report **Field Correspondence**

Date December 31, 1932.

Farm No. 46 Owner A. E. Stevenson

Copied and Sent to Owner

Acres 200

Tenant

Address Heyworth, Illinois.

Crops and Prospect

88 A. Corn Extra good. Total bu. 4,000 - average 45.4.

A. Wheat

5 A. Soy Beans Cut for hay.

80 A. Oats, Barley Total crop 3,500 bu. Owners share.

A. Alfalfa

13 A. Pasture Good - part cut for hay.

7 A. Lots, etc. Small pasture and lots.

A.

A.

2 A. Waste North West Corner cut off by ditch.

A.

A. Total Acres

Clover sown this year. Acres 60 Kind Sweet clover.

Condition Good. 30 South of road excellant growth.

Tenant:

Work Husking corn.

Financial Condition Very.poor.

Spirit and Cooperation Fair.

Weather: Rain and snow - fields very muddy past week.

Improvement to Soil or Bldgs. Since Last Report:
60 acres of sweet clover to be plowed for corn in Spring.

Recommendations, Comments: 30 acres North East of house should be limed if best results are to be obtained.

Signed F. D. Baldwin.

The Decatur Farm Management, Inc.
Decatur, Illinois

Figure 7:5. Farm 46 report. (MCHS)

Figure 7:6. Farm 46 in the Summer of 1997. (W. Walters)

County farmers, one must first understand what happened to the price of a bushel of corn.

When the First World War ended, prices started downward. By 1919, corn had dropped from its wartime high to $1.40 a bushel. Farmers were concerned, but many felt the price would eventually rebound. Concern mounted when, in September of 1920, it had fallen to $1.31. In October, it was down to $0.91. Near panic set in when, in November, 1920 McLean County corn was selling for only $0.71. Most thought the fall was temporary, the result of a good crop, and the fact that railroad cars were now readily available to ship the grain. Everyone was stunned when, in 1921, corn dropped to $0.45. Occasionally, during the next decade, it looked as if the price might recover. The very small 1924 crop briefly pushed prices close to a dollar, but again fell rapidly. By 1932, corn plunged to $0.25 a bushel, and, to many, the corn farmer's situation seemed nearly hopeless. Farmers expected dips in the price of corn to be measured in months; this one lasted twenty years. Grain became so cheap that there was a movement to go back completely to horse farming. New Deal Programs helped a little, but not much. In the late 1930s, corn was still selling for under $0.40 a bushel.

The post-1920 slump in corn prices was reflected in the value of land. In 1903, an acre of prime McLean County farmland was selling for $90. By 1910, it had risen to $175. During World War I, with corn prices at record highs, it had reached $400. Wishful thinkers talked of prime corn land reaching a thousand dollars an acre, and there was a great deal of land speculation. Fortunately, few farmers went in for this kind of gambling. Arthur Moore, the long time editor of the *Pantagraph*, wrote about this brief period of speculation in his delightful book, *The Farmer and the Rest of Us*, saying, ". . . no real farmer has figured in these sales. They have been made by curbstone dealers in deeds and titles. The game is to leave someone else with the land when the boom dissolves" (Moore 1945, 119-120).

Low land prices prevailed for 20 years. By 1930, farms had dropped to only 45 percent of their 1920 value. Between 1930 and 1935, they lost another one third of their remaining value. As land values went down, so did the farmer's ability to borrow and the landlord's incentive to improve the land. The overall result was twenty years with very little investment in agriculture. Farmers borrowed to survive, and the few dollars available went for interest payments. Often it was the banker who made key decisions about the sale of corn. As Merlin Eugene Tyner put it, "your marketing plan was when the banker said, I need some money to pay off your note, you sold your corn" (McFIP).

Money was not the only problem. Nature also launched a massive assault on the corn farmer. The assault was led by of millions of chinch bugs. The chinch bug is one of nature's great oppor-

Figure 7:7. Melvin Hendricks and his wife Ada on their wedding day, February 25 1939. (Melvin Hendricks)

tunists. Originally it was strictly a tropical insect, but when European farmers began clearing large areas on the east coast and planting them in corn, the chinch bugs saw their chance and began moving north. They were first recorded in North America in 1783, tiny creatures about a sixteenth of an inch in length. Young chinch bugs, not yet able to fly, penetrate the leaf surface with a sharp, slender beak and suck small amounts of sap. Often the first clue to their presence is when the farmer notices the leaf begin to turn brown and wilt (Metcalf 1993, 9.19).

Chinch bugs had long been a problem in the Cornbelt. They are individually inoffensive but collectively, massively destructive. Chinch bugs came in great numbers in 1922, devastating the corn crop. In 1923, they returned and did more damage. The nineteen thirties were even worse. Chinch bugs thrive in hot, dry weather and the mid thirties brought the hottest and driest summers ever recorded in McLean County. In 1933, chinch bugs destroyed an estimated 3,000 acres of corn in McLean County (Thompson 1994). In 1934, chinch bugs ruined 40 million dollars' worth of crops in Illinois (Metcalf 1993, 9,19). Evelyn Schwoerer remembers how her father tried to save his crops. " I can see the post holes my father dug with the post hole digger. Every so far, I don't remember how far apart they were, but they were pretty close. You poured black old creosote oil around those holes. The chinch bugs did not fly. They crawled on the ground. They were black. The ground was just black and moving. They would eat the corn up completely. Just eat it up completely, but if you dug enough holes and they couldn't get out so it would kill them, we didn't have chemicals." The next year, grasshoppers came. "I can remember my father getting on a horse and riding through the corn field, and the corn was shoulder high by then, and he would spray — he had a knapsack . . . and he would swing it back and forth and he had mixed sulfur molasses and something else, bran I believe, so it would all cling together, and he would go back and forth and swing it out of his bag on top of the corn. That would kind of put a damper on some of the grasshoppers" (McFIP). Some, but not many. In 1937, the McLean County Farm Bureau supplied farmers with 20,000 pounds of white arsenic in an effort to poison the "hoppers."

Russell Harris remembers the chinch bug invasions. He recalls his father driving the tractor back and forth all day dragging a milk can filled with weights in an effort to crush the tiny pests. He also remembers chinch bug populations so thick they would blacken roads, making them slippery and a danger to traffic. Like Evelyn, Russell recalls using creosote to channel chinch bugs into gasoline filled holes. No one felt sorry when the gasoline was ignited (Personal conversation).

Through all of these tribulations, corn farming continued. Because tenant farms have always been so important in McLean County, one can follow rural problems during the heart of the Depression by looking at farm management reports from one such operation. (Two of the tenant's names have been changed, and the replacement names are noted with * when first mentioned.) Farm 46 was the designation given to 359 acres located near Heyworth and owned by Bloomington's Stevenson family. The land was high quality upland prairie, deep black soil, naturally rich. Farm 46 was a tenant operation and, like many such farms, the tenant was directed by a professional manager. Some of the accounts which these managers provided to the Stevensons have been donated to the McLean County Historical Society, and they provide the basis for the remarks which follow. Farm 46 had been a profitable one and from 1915 through 1917 returned to the Stevensons an average of about $1,500 each year. Earnings increased in 1918 to $3,349. For 1919, they were $8,100. Then came the bad years. In 1920, Farm 46 paid the Stevensons only $872 and in 1921, $410. Reports from the late 1920s are missing, but a 1931 document draws a particularly detailed picture of the farm.

There were three major buildings on Farm 46: a house valued at $1,995, a barn valued at $1,187, and a corn crib valued at $1,235. In addition, there was a summer kitchen, coal shed, chicken house, and farrowing house, each worth less than $250. Mr. Bunder* the tenant was described by the report as "a willing cooperator in all the work we have outlined for the farm, but at the same time he is a man who lacks timeliness as well as good judgment." The characterization continues. "He is a man who delays having his tractor put in shape until it breaks down or runs out of twine or oil in the middle of the day and then has to lose one-half or three-quarters of a day getting supplies." The manager thought Bunder would give satisfaction in a year such as 1931 when the weather caused few delays, but not in a year like 1929 which presented many problems. At the root of the problem lay money. "Mr. Bunder is also very weak financially, and the fact that he is back for considerable cash rent does not seem to worry him in the least, and while I like Mr. Bunder very much I am of the opinion that unless he can arrange to meet his cash payments that it might be best to replace him at the close of the coming season."

Erosion was a serious problem on Farm 46. "Extensive washes of ditches had been allowed to form along the in the draws; especially this was true on the 120 acre tract west of the house." Subsequently, tiles were repaired and temporary dams made of old woven wire and straw were built to control the loss of soil. August had been extremely hot, but the corn crop was good, averaging 44 bushels of very good quality corn to the acre. Heat had badly hurt oats and clover. Beans looked good, but the tenant's turn to use the shared combine had not yet come when rainy weather arrived, and Bunder's soybeans were still unharvested.

The 1932 report shows Bunder still renting Farm 46, but his financial condition had deteriorated. Bunder could not meet his cash rent for 1931. In place of this rent, he had turned over 500 bushels of corn from his share of the corn crop to the Stevensons. The manager found this arrangement "a rather liberal discount," but reflected, "I can not help but feel but we are ahead just that much as with present prices it seems almost impossible for any tenant to pay his cash rent." Sad words indeed, but at least Bunder was a better tenant than one of the renters of another Stevenson holding, Farm 45, located in Indiana. On that farm, the tenant had failed totally, "one main reason for his failure being too much drinking. He and his brother have been using the house as a still for corn whiskey."

On Farm 46, two thirds of the land was in corn, "which is our most profitable farm crop." This corn averaged 45 bushels to the acre,

Figure 7:8. Melvin Hendricks, on the tractor, was one of many young farmers who were enthusiastic about tractors. (Melvin Hendricks)

which the manager felt was quite good, but "the local price of corn and oats is the lowest in history." The Stevensons' share of the corn crop, stored on bins on the farm, was valued by the manager at only 14 cents a bushel. On the plus side, there were 30 acres planted in good clover ready to be plowed under, and the manager felt the rotation system to be a good one. Bunder's total farm income must have been well short of a thousand dollars out of which he would have to pay for all of his farming implements. He was deeply in debt to his landlords, and the manager summed up Bunder's financial condition with two words: "very poor."

Nineteen thirty-three brought no improvement to Farm 46. Because of heavy rains in the preceding year, sweet clover "became very rank." Clover ground had been too wet to plow with horses, so Bunder had hired it done with a tractor. However, the plowing was done much too late. As a result, the expected higher yield for the first crop of corn harvested after clover had been plowed under had not come. Indeed, the 1933 corn crop had been very poor. In 1934, Bunder was gone. The new tenant, Caldwell*, was said to be "showing considerable interest in fixing up about the place, and in livestock, and it is hoped that it will be possible to work together and improve the appearance as well as the earning power of this farm." Unfortunately for the Stevensons, Caldwell also had his problems. He purchased a used tractor in the spring, but it proved something of a "white elephant." Eventually the dealer took the tractor back; by then, much time had been lost. When Caldwell finally left in the spring of 1937, the manager reported that, "the buildings were in need of repair, the fences were down, and Mr. Caldwell has allowed weeds to flourish in every degree. Mr. Caldwell was not satisfactory and was moved for that reason."

The next tenant was Luther Hartwig. Hartwig had been operating Farm 1, near Cerro Gordo, Illinois. He had done such a good job there, that he was given the opportunity of moving onto Farm 46. It proved to be an excellent move for both Hartwig and the Stevenson family. "Considering the way this farm has been operated in the past years, the year of 1939 has been a highly satisfactory one . . . Since he moved to our farm, I have had the opportunity of working with him and studying him very carefully; and I find him to be honest, industrious, a good cooperator, has excellent equipment, and in my opinion is one of the best tenants available." Hartwig quickly set Farm 46 into good order and began a long and profitable relationship with the Stevensons.

In the 1920s, McLean County was still a place where most cornfield labor was done by horses and by human muscle. Well into the 1930s, tractors were still uncommon, and often mistrusted. For most farmers, the real transition from horse to tractor took place between 1935 and 1940. This change can be followed by looking at the early career of Melvin Hendricks who grew up on a farm near Danvers. Melvin's father owned 200 acres and rented another 160, giving him a respectable sized operating unit. However, size counted for little when there was no money to be made from any amount of corn, or for that matter from any other agricultural crop. Other farm slumps had lasted only a few years. This one stretched on for almost a generation. These were dreadful times. Young Hendricks summed

Figure 7:9. Much of the early debate on tractors centered on the question of rubber vs. metal wheels. Both tractor and driver required lubrication. (MCHS)

up his adolescent years when he wrote, "Prices were so low it was all unprofitable, we lived, but scarcely" (Hendricks memoirs).

In many ways, Melvin Hendricks had a nineteenth century boyhood. If there had been a heavy corn crop in the preceding year, his father burned the remaining stalks. This practice, which had for forty years been condemned by professors of agriculture, is a good example of how, even in the nation's foremost corn growing county, textbook recommendations meant little to practical farmers. Before burning, dried stalks were first harrowed and then raked. Then farmers waited for night. Hendricks wrote, "We always burned stalks at night. It was more fun. It was a beautiful sight, it would light up the whole sky. Especially in the hill country. Row on row was burning on the hills. We could see the glare in the sky when others were burning stalks" (Hendricks memoirs).

New machinery had begun to appear, but for most McLean County farmers motive power remained unchanged. Disks had begun to supplement harrows in field preparation. On the Hendricks farm, fields were double disked before planting in order to break up remaining corn stalks, then they were gone over with a four section harrow which smoothed a 16-foot swath. Four horses were needed to pull the disk, hard work for both men and horses, but harrowing was worse. Unlike other farm implements, the driver walked behind a harrow. After a few rounds, the operator was covered with dust and dirt. Melvin would return to the house "almost too tired to eat." It was even worse on the horses. Great raw sores would develop on their shoulders. These would be treated with various ointments and perhaps covered with pads, but neither solution was entirely successful. Planting by check-rower brought out the competitive instinct among farmers. The alignment of corn rows was never a private affair. When the corn came up, everyone in the neighborhood could see how well you had planted, and they were sure to comment to everyone else on the exactness of your rows. There were also hedges. A half a mile of Osage orange fence on the Hendricks farm had to be trimmed three times a year.

In 1932, Melvin Hendricks turned twenty-one. Because farming had been in so much trouble during the 1920s, the shock of the 1929 collapse was particularly devastating to McLean County's city folk. Almost overnight, factories closed. Urban population declined. Young men and women drifted back to their parents' farms. There was food, and, of course, there was always work to do, but there was no money to pay for the work. Hendricks wrote, "Anyone who had a job, farm or city, was considered very fortunate." Melvin was among the fortunate. A neighbor, Everett Yoder, hired him to work on his farm. His pay was $10 for March and $15 for the remainder of

Figure 7:9a. During the first third of the present century much work on corn was still done with horses. (Funk Heritage Trust)

the summer until husking time. For husking, he would be paid one and a quarter cents for each bushel husked. Because Melvin was young enough and skilled enough, he could husk a hundred bushels a day and make $1.25 a day. "This was big money but awfully hard work." To put his wages in perspective, in 1933, Hendricks was making about a quarter of what he would have been paid for husking in 1917 and about a third of what his wages would have been at the turn of the century. In winter, Melvin returned home and husked for his father. Yoder was pleased with Melvin's work. In 1934, his salary was increased to $20 a month and a cent and a half for each bushel husked.

Work was done with mules as well as horses. One day, while cultivating corn with a two row cultivator and a three mule team, Hendricks fell asleep in the middle of a row. The corn was four to six inches tall. Had it been a team of horses, Hendricks recalled, they would have probably continued plodding down the row and the damage would have been slight. Mules were not nearly so well mannered. Melvin's mule team angled across the field, dragging the cultivator behind them and mowed down two rows of young corn. When the cultivator reached the end of the field, Hendricks awoke with a jolt. The young corn eventually recovered, although the rows were never as tall as the ones nearby. The affair did nothing to increase the young man's love of tilling the soil with animals.

It was the custom for a young farm hand to work no more than two years for a single farmer. So, in 1935, Melvin Hendricks hired on with Claude Otto. Otto was a pleasure to work for; moreover he had a Farmall tractor. It was the first tractor in a neighborhood where most of the land was still worked with animals, although was almost twenty years after the great Bloomington tractor show had attracted so much public attention. The gap in time is testimony both to the mechanical problems of many early tractors and to the unevenness of the diffusion process within the county. Moreover, having a tractor did not mean that it replaced horses for all farm tasks. On Otto's farm, planting and husking were still done in the traditional way. For Hendricks, it was love at first sight. Otto rigged an umbrella over the tractor's seat so Hendricks could cultivate in the shade. As the days wore on, other men cultivating by horse soon found their animals winded and quit for the day or gave their animals long rests, but Hendricks kept on cultivating corn. The only incident occurred when, one day, he heard a loud bang and

Figure 7:10. Single row corn picker about 1935. (Funk Heritage Trust)

snap, and turning around, discovered that the bracket holding the umbrella had sheared off, leaving the umbrella at the end of the row and forcing him to work the remainder of the day in sunshine. It was a minor setback. Hendricks decided that for him, the age of horse farming had passed.

Other farmers remember how rapidly horse farming vanished in McLean County. Walter Weinheimer began farming near Lexington in 1935. "I bought a Case tractor in 1938 and I started selling horses. Course I kept one team, well, for about 4-5 years. I kept one team. Finally sold that one. Everything was all horses before that." Like all farmers, Walter remembers his first tractor. "I got one of the early ones on rubber [tires]. Debated for weeks on whether to get rubber or steel. Figured every way I could do it and finally decided on rubber. That was in 1938" (McFIP).

In one respect, the advent of tractors meant more work for farmers. When horses became fatigued, it meant the end of the work day. Merlin Eugene Tyner remembers, "You usually worked most of the daylight hours in the field. After you quit using horses, of course, you could stay as long as you could stand it. When we got tractors we ran 12 to 14 hours a day" (McFIP).

The end of horse farming had come suddenly. In the mid 1930s, horses were still common work animals on McLean County farms. In 1930, there was still only one tractor for every four Cornbelt farms (Hudson 1994, 192). By 1950, 3,228 county farmers responded to the Federal census of agriculture question about the use of horses. Of these, 2,740 used no horse drawn equipment, 491 used horses along with tractors, and only 15 used only horses. It was a rapid change and one with many consequences. One of these was large scale abandonment of oats. which had always been McLean County's second crop. For almost a hundred years, about a quarter of all McLean County farmland had been devoted to this food for horses. There had always been a little wheat, especially in the very early years or occasionally when — as in 1899 — the price of wheat was exceptionally high, but threshing in McLean County most commonly meant threshing oats. Suddenly, the need for these oats was gone. Gone too, was much of the need for grasses which had always been cut for hay. These changes meant that the four crop rotation also was doomed. By 1950, the need for oats to feed farm horses was gone and the demand for hay was greatly diminished. Another crop would have to be found.

Another effect of widespread use of tractors was renewed interest in mechanical corn pickers. Attempts to mechanize corn harvesting go back at least to the 1820s. By 1874, manufacturers were producing roller type corn pickers which also removed most of

Figure 7:11. Soybeans were the miracle crop of the 1920s and 1930s. (Funk Heritage Trust)

the husk (Hurt 1982, 62-63). McLean County farmers tinkered with these machines, but most rejected them. Corn picker adaptation was closely linked to practical, affordable farm tractors, and the real transition in McLean County came in the late 1930s. By 1950, about half of the farms in the county had a mechanical corn picker and roughly the same number had a combine, although the combine was not yet commonly used to harvest corn. Walter Weinheimer recalled the end of hand picking on his farm, "Well, married in 1935, I know I had help the first two years shucking corn. There was a guy working for me in the summer, he kind of got in trouble with a gal. He got married and had to leave. I got a guy out of Kentucky come up here, stayed up here, shucked corn. Next year I had a guy from southern Illinois shuck corn for me. I think the third year I hired Earl Webb with a picker [he picked] the south part of it. I shucked the rest of it myself. . . . Probably '37-'38 along there I started using a picker. I didn't have one of my own but hired it done. Then my brother and I bought one together and we picked corn together. We bought a Case pull-type picker. We picked both places" (McFIP).

The great crop change of the 1930s was the substitution of soybeans for oats in the system of crop rotation. Soybeans were not entirely new to McLean County. Funk's 1903 catalogue had offered soybean seeds, but it was not until the vegetable oil shortages of World War I that serious attention was paid to developing them as a major agricultural crop (Cavanaugh 1959, 348). The first soybean-crushing mill had been constructed in Hull, England, in 1908; its development was closely linked to Imperial politics. After the Russo-Japanese war, Japanese interests dominated southern Manchuria which meant that beans could be shipped from Darien to England for processing (Windish 1981, 11). The potential of the Manchurian crop for both oil and as a protein supplement for animal seed was quickly recognized. Henry Ford, Eugene Staley, and Gene Funk were among early promoters of the crop. McLean County farmers quickly recognized the essential similarities of the Manchurian climate with that of central Illinois. Serious production of the crop would have to wait for expanded demand and for the establishment of processing facilities. Funk was well acquainted with a soybean processing mill which had been built in 1919 by George Britt and I. C. Bradley in Chicago Heights, Illinois. In 1924, a processing plant was established in Bloomington and soybean production increased rapidly (Cavanaugh 1959, 348-351).

Figure 7:12. Funk Soybean Processing mill about 1920. (Funk Heritage Trust)

At first soy beans were planted with a wheat drill, in rowless fields, and harvested by hand. By the late 1930s, Illinois farmers hit upon an elegant solution. They began to apply newly popular mechanized corn producing techniques to beans. They started planting soybeans like corn, in rows. And, after much experimentation, they discovered that, with some adaptations, corn harvesting equipment could be used to gather beans. The Second World War greatly accelerated the demand for soybean products, especially soybean oil. Soybeans replaced oats in the crop rotation system. For those who witnessed this transition, the speed of the change was breathtaking. "Once in a lifetime!" said Gene Funk, "Yes only once in the annals of crop production has our agronomy experienced anything like the soybean — oats fell in acreage before it. Corn and wheat were challenged as cash income crops. Even in the realm of soil building it threatened established legumes" (Cavanaugh 1959, 349). Funk was troubled by the speed with which the soybean revolution had swept over the Cornbelt. Others were equally disturbed. If change could so rapidly sweep away an established crop like oats, what other changes might come with equal speed? If sudden change could so alter crop patterns, could not other institutions vanish with equal speed? No one would miss the long hours of brutish manual labor that had always been the farmer's lot, and not many would miss the bad roads or rural isolation, but many wondered if much that was desirable in rural life might also be swept away by such powerful forces.

Tractors and soybeans may have arrived, but practical corn herbicides were still in the future, and for many years to come, the struggle with weeds was still fought with ancient weapons. No amount of cultivation would remove all weeds. Wally Yoder remembered, "In those days we had no chemical weed killers and . . . that particular farm had a lot of weeds on it. So we walked through the corn field, usually with a hoe, or a corn knife and cut out the weeds, probably three times a season. [It] wasn't too bad the first time or two, but the third time was after the County Fair when corn stalks were tall and starting to dry up and they were just like a knife. I would walk with one hand up over my face to keep those corn leaves from just slashing your face. That wasn't very much fun but we had to go through the third time to cut out the butterprint, the pig weeds and the weeds that were in that corn because we would cultivate it mechanically"

Figure 7:13. Walking corn about 1935. (MCHS)

Figure 7:14. James R. Holbert worked with the Funk Brothers in developing many new plant products. (Funk Heritage Trust)

(McFIP, Wally Yoder). It was not just a question of corn yield. Yoder believes that McLean County has a special pride of ownership. Weed-free roadsides, well maintained roads, and clean corn rows reflect more than economics.

The 1930s mark the arrival of hybrid corn in Illinois. As farmers struggled with the Depression, scientists had continued to probe corn genetics. The fundamental breakthrough came slowly and, in part, accidentally. The key which unlocked the puzzle came from the creation of something known as an inbred. Inbreds were not originally developed in order to produce hybrid corn. Eugene Davenport of the Department of Agriculture at the University of Illinois was interested in increasing the protein content of corn. He hired a chemist, Cyril G. Hopkins, and together they selected 163 ears of high protein corn. Kernels from each ear were planted in a single row. These were forced to self-pollinate; in corn breeder's jargon, they had been "selfed." In 1900, Edward Murray East was brought in to supervise the project. East found that descendants of one ear consistently had the highest protein content. More important, each self-fertilized generation produced higher levels of protein. There were undesirable side effects. Each self-fertilized generation was smaller, less vigorous and provided lower yields than its parents. Casual undergraduates at the University of Illinois who watched the corn experiments from 1900 to 1904, must have been amused. In four years of tinkering, the scientists had managed to produce the worst corn they had ever seen. In fact, the scientists had done something very important. They had reduced the ears to their lowest common denominator and created an inbred (Crabb 1992, 22-23).

While scientifically interesting, no one predicted a commercial future for inbreds. Hopkins decided inbreds had no value, but did not order the experimental plot abandoned. Moreover, corn from the plots was stored in the Agricultural Experiment Station's grain vaults. In August 1905, Edward Murray East left Illinois for the University of Connecticut, but he persuaded a friend to send him a few kernels of selected ears which had been harvested in 1905. In the spring of 1906, these were planted in Connecticut. East's plan was to "self" the resulting corn one more time and then use these refined inbreds as male parents to cross with the leading varieties of open pollinated Connecticut corn. He then began trading ideas with pioneer corn breeder George H. Shull. The breakthrough came in 1908. By crossing two inbreds, Burr and Leaming, he produced a single-cross hybrid with remarkable yield and astonishing uniformity. Always the good scientist, East repeated his experiments and obtained the same results. He had produced the first true hybrid corn (Crabb 1993, 26-27).

The road from the tiny plots of college experiment stations to the fields of McLean County was long and difficult. Among other

Figure 7:15. The development of grassed waterways in the late 1930s revolutionized the appearance of the Cornbelt. (W. Walters)

things, in 1908, the theoretical genetics associated with hybrid vigor in corn were not well understood. Until they were worked out, corn breeding could never be done on more than a hit-and-miss basis. It took a great scientific effort in the teens and twenties to develop genetic theory to the point where corn breeding was predictable (Hayes 1963, 41-81). In addition, what East had done with his single crosses was only one of many ways in which corn could be manipulated to produce the desired results. On a larger scale, the problems were equally complex. At first, there were only tiny amounts of the new inbreds and hybrids available. Translating these into a product which fit the particular needs of farmers and which could be produced in vast quantities with consistent quality was a huge task. Explaining these hybrids to those who would have to be persuaded to plant them and justifying to Depression-weary farmers, the need to pay for such expensive seed was an equally daunting task. In these latter endeavors, central Illinois would have a special role to play.

Gene Funk and his still-young seed company were intensely interested in corn hybrids. The company had excellent relations with farmers all over the county, a reputation for quality, and more importantly, a tradition of serving as a center for the exchange and distribution of corn information. Moreover, Funk Farms was the site of a Federal Field Experiment station. Gene Funk and James R. Holbert both had a deep, personal interest in developing new strains of inbreds. By the early 1920s, work on commercial hybrids was rapidly advancing. As Richard Crabb wrote, "As a result of Holbert's own work and his willingness to accept inbred material from others in exchange for his own, the first outstanding hybrids in the Cornbelt grew on Funk Farms" (Crabb 1993, 131). By 1923, they had produced tribrids, that is, hybrids from three inbreds, which seemed to have considerable commercial potential. In 1925, they had successfully created double-cross hybrids, corn with four sets of inbred parents. That was enough to convince Gene Funk that the time had come to permit farmers to start producing some of the new corn. The G. J. Mecherle farm nine miles east of Bloomington, operated by Walter Meers, was one of the most active of these early farm test stations (Crabb 1993, 129-134).

Farmers were often skeptical of new products and hybrid corn was no exception. McLean County farm writer H. Clay Tate recalled a conversation with hybrid corn pioneer, James R. Holbert, who remarked, "You almost had to pay farmers to plant hybrid corn" (Tate 1972, 116). The experiences of Carl Graff, Sr. illustrate the reluctance of some farmers to accept what researchers and salesmen assured them was true. The elder Graff was born in 1904 near the lovely Bavarian town of Passau, where the Inn flowing north out of the Alps joins the Danube in the southeastern corner of Germany. When his father and uncle were drafted into the army,

Figure 7:16. This picture in *Life* aroused the anger of many Midwestern farmers. (*Life*)

Carl quit school to work on the family farm three miles from the Danube. Wheat and barley were hidden inside the walls of buildings to keep it from being confiscated by the army, but solders took all of the livestock except for one horse, so weak it could barely walk. After the war, Carl worked on state land as a forester and tried to eke out a living selling firewood, but there was little money to be made in rural Germany in the 1920s. At last, Carl's uncle who lived at Covell in McLean County agreed to sponsor his nephew's immigration. Carl tucked the three or four dollars his uncle had sent him into his pocket, someone pinned a note to his shirt, and he set off for the new world (McFIP Carl Graff, Jr.).

Graff owed his uncle $300 for passage money. He cut weeds and he husked corn, but it took three years to pay back the passage money. He took any work he could find, harvesting wheat in North Dakota, working as a waiter, washing dishes in a hospital, and then back to McLean County to work on his uncle's farm. At first, he knew nothing about corn or soybeans, but he understood livestock and was ready to learn the rest. At that time, all of the corn grown on the farm was open pollinated, but there was talk about the new hybrid varieties. A seed salesman promised Carl that if the new hybrid seed didn't grow more corn, he would give Carl the seed. Carl came home with a bushel. "His uncle thought it was a waste of money at $6 a bushel when corn was worth about $15 a bushel" (McFIP).

Using a two row planter, Carl put in fifteen or twenty acres of the new seed before the supply began to run low. He then filled one of the seed boxes on the planter with open pollinated corn and the other with what remained of the new hybrid. His uncle was skeptical and his skepticism increased early in the season when the open pollinated row grew more rapidly. Hybrid corn, the uncle insisted, was no good. Carl made a bet with his uncle. "I will shuck out my row and you can have the other row" (McFIP). By the time the corn had tasseled, the hybrid corn had caught up with the open pollinated row. When ears formed, it was clear to both men that the ears in the hybrid row were both larger and more even. The shucking contest never took place. A summer storm swept over the cornfield

Figure 7:17. Laying tile in eastern McLean County about 1940. (MCHS)

and toppled the open pollinated row while the hybrid row remained upright. The next year the uncle joined the ranks of farmers planting hybrid corn (McFIP, Carl Graff, Jr.).

The year 1936 marked the real start of the hybrid corn revolution in McLean County. Until then, seed salesmen had been frustrated by their inability to get more than a few farmers to purchase the new varieties of seed. It was indeed a revolutionary and effective product, but most farmers balked at the eight to sixteen dollars a bushel price. Nature came to the rescue of the struggling seed companies. For corn generally, 1935 was a dreadful year. In addition to all of the other problems, root worms destroyed thousands of acres of corn, but those fields planted with the new seeds came through with record yields. Nothing spurs a McLean County farmer into action faster than the sight of a neighbor's field looking better than his own. In 1936, sales of hybrid seed soared. It was another bad year for corn, one of the hottest and driest ever recorded, but the hybrid fields came through with flying colors. The rush for hybrid corn seed was on. It was a bonanza for the seed companies. Hybrid vigor lasted for only one crop; if planted, seed from a hybrid ear was no better than that of its open pollinated ancestors. From a financial standpoint, hybrid corn has one great advantage. Each year the farmers had to return to the company for new seed. Much of the money coming into seed company treasuries went for research into developing new and better hybrids. The race for better corn which would transform the Cornbelt had begun. In 1959, 94.8 percent of the corn planted in the United States was hybrid

Figure 7:18. Corn field in eastern McLean county about 1920. Wide rows and low plant densities still prevail. (Funk Heritage Trust)

(Hayes 1963, 6). Bloomington became the hybrid seed corn capital of the world. By the 1960s, almost 20 percent of the world's total supply was being grown within a few miles of the city (Tate 1972, 116-17).

Until the 1930s, the Federal government played almost no role in corn belt farming. From the start, farmers had accepted the need for local governments. They acted as road viewers, fence arbitrators, and sometimes reluctantly accepted the need to assess each other's land for the construction of roads and payment of teachers, but this was local, not Federal government. This changed abruptly in 1933. Roosevelt had promised a New Deal and his first administration set out to provide it.

Cornbelt politics have always been pragmatic and McLean County farmers have never been comfortable with demands for social reorganization. To put it another way, they have traditionally feared government bureaucracy more than they have feared market forces, but by 1933, it was clear to many of these pragmatists that market forces had repeatedly and continually failed to provide a price for corn that would permit their survival. Some farmers never accepted government intervention. One recently told an interviewer, "My Granddad, Dad and I had nothing to do with the government. We didn't believe in it, we didn't like them, we didn't want them around . . . government stuff I throw in the garbage . . . I don't want nothing to do with them, and I don't want none of their damn money!" (McFIP). Still, McLean County farmers voted strongly for Roosevelt in 1932 and seemed ready for a change.

The first and most important of Roosevelt's farm programs was the Agricultural Adjustment Act. The idea behind AAA was to establish a fixed level, called "parity," based on prices for corn in the good years between 1910-1914, and to support these prices with payments for agricultural processors. More than eighty per cent of McLean County farmers eventually accepted checks under the provisions of Agricultural Adjustment Acts (Moore 1945,137). After the Supreme Court ruled the acts unconstitutional, farmers

Figure 7:19. Detasselers during World War II. (MCHS)

participated in equal measure in the replacement programs which followed, especially the Soil Conservation and Domestic Allotment Act. Under this act and related legislation, farmers were paid directly by the Federal government for soil conservation measures.

Because of these acts, a new landscape element, the grassed waterway, appeared in McLean County. Today they are so common that they almost never inspire comment, but one must realize that before the Roosevelt administration, there were no really significant national programs in effect to control erosion. Grassed waterways now exist on almost every farm in the county and they have been extremely successful in controlling soil erosion. Farm 46 records the results of this government program. Dams built in 1930 were just a temporary measure. In June 1940, the manager reported, "we were finally able to get the ditches graded shut on this farm, and they have all been seeded to grass for grass water-way." The $358.94 income listed as "Gov. Paymts.," the first of any kind received by Farm 46, may have been for these conservation measures.

One government program which farmers universally applauded was the Rural Electrification Administration. Farm families sometimes remember that the old food tasted better, they recall the old communities as friendlier places, or think that the kids of an older era had more stamina, but none want to go back to the days of pre-electric farming. Government sponsored rural electrification programs came late to McLean County. Many farms were still not electrified in the mid 1930s. Carl Graf, Jr. recalls that electricity came to his farm on Christmas Day, 1940. That summer his father had everything wired for electricity — refrigerator, radio, toaster, and water pump. However, not until December was the power turned on. Running a wire down to the corn crib was one of the first things done. Carl's father hooked up a used electric motor and they began grinding grain with electric power (McFIP).

By 1939, another war in Europe had again begun to change the face of the McLean County corn farm. The most evident change was a rapid increase in the price of corn and a corresponding jump in farm value. In 1928-1939, corn sold for an average of 58 cents a bushel. In 1944, the price was a dollar and seven cents, and farm values were up an average of 42 percent. This increase did not mean that the average farmer was 42 percent better off than he was six years earlier, but rising farm values increased the attractiveness of renting one's farm to a tenant. Arthur Moore explained it this way as he tried to help non-farm readers understand some of the complexities of farming in central Illinois: "Farmers moved to the villages and towns when prices started up because of the war. What happened was that many farmers found a tenant as the higher prices promised them a comfortable income in retirement. The farm which was supporting one family was now supporting two. Is the farm itself gaining anything as a food producer? On the contrary, it is likely to be losing for it must now support both tenant

"Time and Tassels Wait for no Man!"

The peak period is about on us, and within the next two weeks every hybrid seed corn field will required working. Each field must be detasseled to 8 to 12 times for a thorough job. Fields must be covered often to prevent seed parent tassels shedding pollen before they are removed. One day can mean the difference of thousands of bushels fo corn and likewise thousands of pounds of beef, pork, butter and dozens of eggs.

Every hour and Every Helper Counts

Stand by to fulfill your pledge. Be there when help is called. If you have registered you will be notified by mail or by phone when and where to report. All applicants will not be required at once. Await your notice before reporting for work.

Let's Fight the WAR and on the HOME FRONT Like Our Boys Are Fighting on the BATTLEFRONT. Make Your Soldier Proud of You!

Figure 7:20. Appeal for detasslers in the summer of 1943 (type reset for clarity). *(Pantagraph)*

and retired owner" (Moore 1945, 73). By 1940, almost 60 percent of McLean County farms were operated by tenants.

Tenants, as well as landlords, profited from improved prices. On Farm 46, concrete walks were built around the house, and storm windows were added to its north and west sides. The corn crib was straightened and re-roofed. The farm was wired for electricity. "Tenant seems to be very well pleased that we are making these added improvements."

Cornbelt farmers were well primed for the struggle with the Axis powers. Hybrid corn arrived just in time to meet wartime demands. New harvesting equipment meant that farm labor could be freed up for other tasks. Moreover, government policies gave a high priority to agricultural machinery. Conservation measures, started in the 1930s, left Cornbelt soil in better shape than it had been for years. Even the weather cooperated. The 1942 corn crop was the largest ever recorded, over three billion bushels; the 1943 crop was only slightly smaller. The magnificent performance of American agriculture is one of the usually untold stories of World War II.

Even so, there were problems. Mindful of rapid increases during World War I, Congress decided to set ceiling prices on corn. The initial ceiling was set at $1.07 a bushel. As demand rose, farmers fed corn to livestock rather than shipping it directly to market. Because the price of hogs was fixed at a fairly liberal $13.75 per 100 pounds, corn vanished from the commercial grain market and went into feeding hogs. Between 1942 and 1943, the number of hogs on American farms more than doubled and by the summer of 1943, many grain elevators were empty. This situation provided the background for what became one of the most famous farm pictures ever taken in McLean County, a picture which enraged McLean County farmers.

In July of 1943, *Life Magazine*, in an effort to call attention to the crisis, sent a photographer down to Frank Hubert's farm near Heyworth. He took several pictures, including one of bulging corn cribs, but the shot which *Life* selected as the lead picture for its first article showed Hubert's pigs laying in their pen with corn kernels scattered everywhere. The caption was equally inflammatory. "Fat-sated hogs on Illinois farm wallow in thick carpet of shelled corn. Farmers prefer to waste corn on livestock rather than sell it at ceiling prices." The article went on: "Washington's economic contraption, patched together with as many complications as one of Rube Goldberg's machines, has broken down completely" ("Corn," *Life*, 19 July 1943, 23). As Arthur Moore pointed out, within minutes of the picture being taken, the corn had all been eaten and was in no sense wasted. Besides, he sniffed, the hogs weren't fat-sated but young, "80- and 90-pound devils, frisky as hungry puppies" (Moore 1945, 6). Byron D. Kline, President of the McLean County Farm Bureau fired off a letter of protest, but the damage had been done and farmers were angry.

Another nationally circulated picture had a different effect. Mechanized picking had greatly reduced the demand for huskers, but greatly increased use of hybrid seed created a huge demand for detasslers. Detasseling for seed corn production became a critical wartime activity. Businessmen gave employees time off to detassel. On Sunday July 11, 1943, Reverend Loyal M.Thompson of the First Methodist Church preached his sermon on the topic of detasseling and declared, "There is an agricultural emergency upon us." Huge advertisements proclaimed, "Time and Tassels wait for no Man; Stand by to Fulfill your Pledge; Let's fight the WAR on the *home* front like our boys are Fighting at the BATTLEFRONT; *Make your soldier proud of you; Be Ready to Go When Called*" (*Pantagraph* 14 July, 1943, 10). To help promote detasseling efforts, Dr. J.R. Holbert had, in 1942, taken and circulated a picture of Bettie Lou Geneva working with a detasseling crew. A local paper published the photograph.

In the summer of 1943, the Army Engineer Corps at Camp Maxey, Texas, held a contest for "Sweetheart" of the camp. Private Tony Zaffiri didn't have a photograph to submit, so he submitted the newspaper clipping of Bettie Lou. She was elected. Nationwide press picked up the story and it was circulated to hundreds of papers. Letters by the dozen began rolling in and the Bloomington girl in the cornfield was a celebrity. She took it calmly, but surely she had tongue in cheek when she told a reporter. "It's an easy task. Just like walking up and down that's all." Bettle Lou recommended slacks, hat, and a long sleeved blouse as proper detasseling atire. She was anxious for the 1943 season to begin. "It makes you feel you're really going to do something useful, and you can make good money while you are doing it" (*Pantagraph*, 11 July, 1943, 11).

By the time soldiers returned from the war, McLean County farms had undergone a complete transformation. Corn farming was once again profitable. Horses were vanishing. Picking had become mechanized. Hybrid corn was nearly universal. Soybeans had replaced oats. The Federal government had dealt itself into the game of corn farming. Yet there were constants. As before, each year saw farms become a little larger and each year, owner operators working only their own land became a little less common. By 1950, farmers could shake their heads and explain that they had witnessed the greatest transformation ever seen in the Cornbelt. They were right and they were wrong. Yes, the changes had been profound, and yes, they had been amazingly rapid, but, in fact, even greater and much more rapid changes lay just ahead. *Zea Maize* had just begun to release its surprises.

Chapter VIII

YEARS OF CHANGE
1950-1997

In the second half of the twentieth century, corn farming changed more than it had in all previous years combined. All of these changes were linked in so many ways that it is difficult to tell the story of one development unless it is kept in mind that a single change often triggered others. Perhaps, then, it is best to begin with those things which did not change. One of these is the county's dominant crop, corn. Today, McLean County holds its position as the nation's leading corn producing county. When asked what he planted, Ray Wesley Rafferty replied, "well in the last few years, corn and soybeans have been our major crop." He continued, "Corn is our main money-making [crop]. I mean we make more money off of corn than any other crop" (McFIP). Most county farmers will say the same thing. Since 1950, corn has continued to dominate McLean County statistics. Wheat was grown and, of course, beans. Purebred cattle were raised, especially by farmers who didn't need to rely on them to pay bills. There were successful hog farmers. Yet, for most, it was corn that paid for combines, herbicides, and prom dresses. Scott Hoeft is typical. He gets 60 percent of his income from corn. "It is the main crop. Its the thing that everybody talks about. If you go down to the coffee shop, you talk about bean yields a little, but everybody wants to know what the corn is making" (McFIP).

The soil is another constant. Technology is indeed splendid, but it cannot create profitable farmland. McLean County farmers have been willing to spend the huge amounts needed for the develop-

Figure 8:1. By the 1950s trucks completely dominated farm to elevator journeys in McLean County. *(Novartis)*

Figure 8:2. Corn yields increased rapidly in the 1950s and 1960s. Even high-rise reinforced concrete elevators can not always handle the crop. (*Novartis*)

ments described in this chapter only because they have an unshakable belief that this is the finest corn ground anywhere in the world. They were willing to gamble because they knew McLean County soils stacked the odds in their favor. That is, it has always done so for farmers who were ready to embrace change.

During the years between 1950 and 1997, there was an explosive national increase in corn yields. Farmers were amazed when, in 1958, and for the first time in its history, the United States produced over four billion bushels of corn. This feat was particularly remarkable because the acres planted in corn had been steadily dropping. Since then, yields continued to go up. The 1970 corn crop was 5.15 billion bushels. In 1976, the harvest for the first time topped six billion bushels. In 1978, it went over seven billion bushels and by 1981, it climbed to over eight billion bushels. By 1992, it was over 9.4 billion bushels (Crabb 1992, 6-7). The change in corn yields during this period was one of the most rapid and profound in the history of human agriculture. The rapidity with which corn yields increased exceeded that in the celebrated "Agricultural Revolution" of the seventeenth century.

The fact that this much corn could be sold reflects massive changes in world agricultural markets. Much of the corn was destined for rapid export. Like it or not, Cornbelt counties became players in a complex international chess game in which the chess pieces had names like hunger, regulation, boycott, port efficiency, and Brazilian rainfall. Globalization and yield explosion were key components of the story, but there were many subplots. McLean County participated in all of the changes which transformed corn farming. If one seeks an answer to the question, "how were these revolutions accomplished?," there is no better way to do it than to listen to McLean County farmers themselves.

In 1950, McLean County was oddly balanced between the present and the future. Farms were still fenced, and many fields were still completely surrounded with hedges of Osage orange. Most farmers still raised cattle and hogs to supplement corn and soybeans. Chicken houses were common and barnyard fowl frequently ran free. Almost all of the corn planted was hybrid supplied by Funk, Pfister, or one of the other local companies. Just about everyone had a tractor. Each fall, one of the tractor's jobs would be to obliterate all traces of that year's crop. Land was disked and then turned under with a moldboard plow. In spring, before planting, the soil was again worked. It was the goal of each farmer to start the spring planting season with a perfectly level, finely pulverized black field in which not a trace of plant remnant could be seen. The tractor was used to cultivate the growing corn as many as three times in a season to help to control weeds, but all farm children were still expected to know what to do with a hoe in a cornfield. A great deal of manure was applied to fields, but the use of commercial fertilizer was rapidly increasing. Almost all crops were harvested mechanically, but the machinery was very different from that used today. Looking at corn harvesting is a good place to begin the story of the change.

The census of agriculture taken in 1954 shows a county where there were still some survivors of the old system. Just over 10 per-

Figure 8:3. Chisel plowing through 'trash' from last year's crop often begins the farming year on today's minimum till farms. A 45 Cat Challenger pulls a Landall plow. (Ann Charback)

cent of the farms were still being worked only with horses and about 17 percent had no electricity. Eighteen percent still had no running water. On the other hand, two-thirds of the farms had a combine and about half had a television set. The amount of oats being grown was falling, but there were 10 acres of oats for every six of soybeans. Within ten years, these residuals of the old system had vanished. Electricity and running water were nearly universal. Horse farming had ended; in 1964, there was an average of almost three tractors per farm. In that year, there were nearly five acres of soybeans for each acre of oats. These figures, however, do not reflect only improved farming and better economic conditions. In the ten years between 1954 and 1964, one quarter of the farms in the county disappeared, and the average farm size jumped from 204 to over 260 acres.

The miracle machine of the 1930s and 1940s was the mechanical corn picker. In the course of about ten years it had virtually eliminated the crews of huskers who used to migrate northward from Kentucky, Tennessee, and southern Illinois. There was a wide variety of corn pickers in use, and all had serious problems. They were mechanically unreliable. In theory, a farmer should have been able to pick 15 to 20 acres of corn in a day, but that rarely happened. "The equipment is so much better than what it was back when I started. If you went into the field with a corn picker you would probably break down sometime during the day. Anymore we don't ever have a breakdown. Very, very seldom . . . You can't afford to be broke down now . . . a day or even an hour is pretty important at times" (McFIP, Ray Wesley Rafferty). Improvements came slowly. Originally, the gears on corn pickers were exposed. Later, they were encased and run in oil, helping reliability. Corn pickers were also frightfully dangerous. Even in a well-cultivated field, weeds remained, and these tended to jam the picker, or it might jam on a branch. A jacket might drop from the tractor and clog the corn picker's rollers. Early in the day, farmers were careful, but when fatigue set in, or when the stupid picker jammed for the tenth time, or when there was just half a row to finish before dusk, then there was frequently trouble. If power were not be fully disengaged, or a lever bumped, or a companion inattentive; then the farmer might learn that the rollers on a corn picker were infinitely faster than human reflexes. With brutal swiftness, the machines mangled hands and snapped off arms.

There were a number of ways corn pickers could be linked with tractors. Probably the most common method was to pull the picker behind and to one side of the tractor. This setup meant that as the farmer circled the field, he could not harvest the outermost rows without crushing them with the tractor tires. The usual solution to this was to "open" a field by harvesting the pair of rows nearest the fence by hand. Opening fields was the last significant survival of old harvest technology in general use in McLean County. Unlike hand harvesting, the early corn pickers usually delivered ears of corn to the wagon with their husks in place resulting in corn that

Figure 8:4. Sprayers have replaced much of the cultivation once done on McLean County farms. McLean County Farm Service spraying herbicide on land farmed by Murphy Farms in the spring of 1995. (Ann Charback)

Figure 8:5. Harold Smith prepares a John Deere 16 row planter in the spring of 1995. The tractor is a Case IA 7220. (Ann Charback)

Figure 8:6. Darrell Beehn, Jr. loads Pioneer seed into an International 400 Cycle planter near Shirley, Illinois. Darrell Beehn, Sr. stands beside the planter. (Ann Charback)

was slower to dry in the cribs. A logical answer was to add a device which husked, or perhaps husked and shelled the corn at the time it was picked. Such combined machines, or combines, had been used in McLean County for small grains since 1926, but there were problems in adapting combines to corn. For a hundred years, corn farmer's operations had been geared toward the storage of ear corn and his buildings had been developed for that system of farming. Shelled corn was much more difficult to keep for long periods than ear corn. Changing meant spending money and at first, the rate of change was slow.

Farmers developed many ingenious ways of coping with the picking and shelling problem. For example, in 1955, Lester Fuller had a Minneapolis-Moline corn sheller mounted on the rear end of a Massey-Harris corn picker. An odd-looking combination, but it worked well. Stripped cobs came flying out the back end, and shelled corn went cascading into a wagon driven alongside the tractor (*Pantagraph* 13 Oct. 1955, 32). Other farmers worked on similar expedients. McLean County farmers love to tinker, but usually tinker with some serious goal in mind; county farmers count among their numbers many highly skilled mechanics.

The use of the tractor had also altered the way corn was planted. Corn had been planted in what were called hills, although they were usually no higher than the surrounding land. Each hill had about three kernels, and the hills were spaced 40 inches apart at each corner in a checkerboard pattern. Forty inches was needed because it was the width of a horse, and a horse had to pass the stalks in each direction when the corn was cross cultivated to control weeds. In other words, not much had changed since 1850. In the 1950s, many farmers began to think that 40 inches no longer made sense. New hybrids, aided by the new fertilizers, could support more corn plants, but a farmer couldn't add more plants without sacrificing the 40-inch squares needed for cultivation on all sides of the hill. Gradually a few farmers took the risk and increased density by planting kernels one at a time, close together, in long rows. Cultivation could now only be done between rows, but unlike horses, tractors could straddle corn rows without crushing the corn. Therefore it wasn't long before the rows began creeping closer together. Today, 30-inch rows are common and some farmers are experimenting with narrower widths. Narrow rows presented problems for weed control, but they did help boost production. The visual transformation of the cornfield was striking. The traditional cornfield was a sparsely planted mass; the modern version is a series of tightly planted rows.

Retired McLean County farmer, Merlin Tyner, explained the change in row width with these words, "we used to have 38- and 40-inch rows. When I quit I was planting 36-inch rows. Now peo-

Figure 8:7. Narrow rows and high plant density characterize modern corn planting. Contrast this picture with Figure 7:18. (Ann Charback)

Figure 8:8. Mature corn on the Funk Trust Farm near Shirley, Illinois in September 1995. (Ann Charback)

Figure 8:9. Darrell Beehn harvesting with a John Deere 4400 combine near Shirley, Illinois, in September 1995. (Ann Charback)

ple are pretty much going to 30-inch rows because we plant corn almost twice as thick as I planted it when I first started. . . . If you put 27 or 30 thousand plants in a 38-inch row, you have got it almost solid. Where if you put it in 30-inch rows, you can spread out the kernels per row and get better utilization of the fertilizer, more root room and so forth" (McFIP). The problem was that 36-inch rows did not fit well into a world of 40-inch machinery. During long, cold winters, farmers pondered this problem and began to wonder if the existing machines couldn't be adjusted. One of these farmers was Jim Forsyth. "Jimmy also had narrow rows and we had to plow them with a mule and one plow shovel. That's all we had. I mean that is the only way we could do it, because there was no equipment in those days, '39 and '40, to do it. He had already double rowed [planted two rows at a time]. He took his four row planter and double rowed it — went in between. He tried that long before any of this narrow row ever started. No equipment or nothing; he couldn't do that because it was time consuming and he had no equipment to pick it. The only way to pick it was shuck by hand" (McFIP, Arthur Spencer). By the 1950s and 1960s, farm machinery manufacturers began to catch up with innovative farmers and started to build machinery for narrower rows.

At the same time cornfield geometry was changing and rows were becoming narrower, there was an equally important development in corn fertilizing. Corn is a grass and like all grasses responds quickly to increased nitrogen in the soil. For seventy years, it had been known that if one could get the right sort of nitrogen to the corn at the right time the corn would flourish. The problem was that the only effective way to get nitrogen into the corn-growing soil was to plant clover and plow it under. It worked, but the amount of nitrogen delivered was small. The demand for explosives in World War II increased America's capacity to produce various nitrogen compounds. After the war, manufacturers tried to use some of this excess wartime capacity to produce nitrogen based fertilizers. There were many technical problems, but for the farmer, the major problem was cost. Ultimately, the most cost effective products often proved to be one of the various forms of ammonia, in particular, anhydrous ammonia. It took many years of trial and error before the application of this product became a standard procedure on Cornbelt farms.

About 1950 anhydrous experimenters first brought their product to McLean County. Ray Wesley Rafferty remembers that he first applied anhydrous in 1951, "We used anhydrous ever since they came out with it. It is a nitrogen that really boosted the yield of corn a bunch." Then ammonia cost $60 a ton; in 1996, it was about $450 a ton. "Back when we first started farming," Rafferty continued, "there is no way we could afford to do that. The only way that you can do it now is to farm a lot of land and we farm so much more than we used to" (McFIP). Merlin Eugene Tyner remembers when

Figure 8:10. The corn head of a 9600 John Deere Combine during the 1995 harvest. (Ann Charback)

anhydrous ammonia first came to McLean County. "Yes there was a group of guys came through here out of Arkansas. They rigged up . . . a Massey Harris tractor and they had put knives on it to apply anhydrous. And, it had a tank up behind the seat. I think it held 200 gallon or something like that, 250 maybe. Then they'd get anhydrous into Stanford on a railroad car and they had a truck that they would haul it from the railroad car out to the machine." They put on 60 pounds of anhydrous, the equivalent of about 48 pounds of nitrogen (McFIP).

Like all other Cornbelt machinery, ammonia applicators grew in size. Soon, Bill Wiltham was using 1,000-gallon trucks to supply five-knife machines equipped with their own 500-gallon tanks which were able to cover 60 acres a day. Farmers were impressed. Today, 20 to 30 acres an hour is not uncommon. Anhydrous is only one of many fertilizers extensively applied. Fertilizers came to be used in conjunction with rigorous programs of soil testing, with the farmer often attempting to replace the exact amount of nutrients taken out of the soil by last year's crop. It is a matter of both soil chemistry and careful cost control. All such products are expensive, and the amount of fertilizer applied can be impressive. One typical McLean County farmer reported that in 1996 he had applied 250 pounds of Dap, 250 pounds of potash, 150 pounds of anhydrous, and two tons of lime to his corn.

Anhydrous and the other fertilizers brought their own problems. Weeds enjoyed the nitrogen almost as much as the corn did and the problem was compounded by the end of cross cultivation. Chemical herbicides provide an answer. At first these were applied as limited supplements to cultivation, but by the 1960s, farmers increasingly began to see herbicides as replacements for cultivation. There have been complaints about their use. Arthur Spencer talks about the negative influence of what he calls the "do-gooders." "If we would go back and farm the way my granddad did, my dad did, three-fourths of the people in the United States would be starving to death. Because, we couldn't control the weeds to raise that much corn. When I was a kid, we had to cultivate the corn four times" (McFIP). Part of the weed problem today is that corn frequently follows corn on the same field. In traditional four year rotations, the break from corn — when land was in oats or pasture — helped to control weeds. This benefit is no longer present, the usual answer has been more herbicides.

Chemical weed control eliminated almost all hand cultivation of corn. Not many regret the end of hand cultivation. "I can remember we would walk cornfields and take out the weeds after the corn had tasseled even, and it was hot, hot work. When we speak of the good old days we kind of forget stuff like that" (McFIP, Donald Pleines). Changing weed control has made moldboard

Figure 8:11. Darrell Beehn, Jr. adjusts a belt on a John Deere 4400 combine in October 1995. Absolute mechanical reliability of harvest equipment is a high priority for McLean County farmers. (Ann Charback)

plowing, that is, plowing where the soil is turned over, rare. The harrow has vanished from McLean County farms. Today, plowing to control weeds is often restricted to narrow chisel blades which cut and lift the soil but do not turn it over. The war on weeds has become dominated by chemicals, and it is an expensive war. Consider these 1996 prices paid by one McLean County farmer. Before the corn came up, there was an application of Top Notch at 19 dollars an acre. After the corn emerged, it was sprayed with Laddock costing nine dollars an acre. For a thousand acres of corn, this works out to an expenditure of $28,000 just for these chemicals. Added to this are the cost of additional weed killers and spot spraying for problem areas.

In its extreme form, un-plowed and un-cultivated corn is called no-till. The term is a little hard to define because farmers practice many different degrees of no-till. Basically, this farming method involves substituting chemical weed controls for cultivation. It also involves not removing or plowing under the accumulated stalk and cob debris left by harvesting, but leaving all or most of them to lie like mulch on the surface. In part, no-till was designed to prevent soil compaction by limiting the number of times machinery had to pass over the cornfield. Its more important function was soil conservation. No-till imposed a barrier of dead organic matter between the soil and the atmosphere. Raindrops were absorbed by this barrier, and wind was less able to reach and dislodge individual soil particles. Evaporation is retarded and the reflection of sunlight is increased. Because of this organic litter, the germination of seeds is delayed, pesticides remain active for longer periods and root growth may be slowed (Gersmehl 1978,66). No-till controls wind erosion, water erosion, and weed growth. Ever since its introduction in the mid 1960s, no-till has been growing in popularity. A modified form of no-till, minimum-till, where only some organic litter is left on the surface, has been particularly popular in the central Cornbelt. Ken Dunahee says, ". . . especially if you have a little bit of rolling ground. It sure protects the ground from washing away. Right around this area I would say there is more minimum-till. They do just a small amount. They try to leave a lot of stalk[s] on top to keep it from washing and blowing. Sure got away from that clean plowing they used years ago" (McFIP). One still sees plows at work, but their numbers are diminished. "No, about one trip over in the spring and plant it. Just enough to level it out and leave some trash on top and then they'd plant it" (McFIP, Ken Dunahee). Some still disk in the fall to work stalk remnants into the soil, but far less disking is done than in the past. Today 17 percent of McLean County corn is grown by no-till and another 22 percent by minimum till methods.

No-till/minimum-till leaves fields with a "trashy" appearance which some old time farmers find disturbing. Generally, the new techniques involve less time spent in field labor and more time spent in management. Fuel costs are reduced but pesticide costs

Figure 8:12. Harvesting corn on the Funk Trust Farm near Shirley, Illinois, October 1995. (Ann Charback)

are increased. Some McLean County farmers are convinced that these procedures have increased problems associated with cornfield insects. Others argue that no-till can be a risky procedure with corn; sensitive high yielding hybrids may not be the ideal plants for this kind of tillage.

Just after World War II, DDT began to be widely used as a farm insecticide. Some farmers still regard it as one of the most effective farm insect control agents they have ever used; certainly it worked better than crushing insects with a weighted milk can. Health concerns were soon raised about DDT. Thus began a cycle which was repeated several times: an insecticide would be introduced, prove its effectiveness, and then be withdrawn or restricted. Changes in insecticides have been even more rapid than most changes in Cornbelt farming. Almost all farmers use some insecticides, and the recommendations seem to some to change almost monthly. It is a three sided battle among the inventiveness of farm chemical engineers, the restrictions placed on farmers by safety-conscious government regulators, and the adaptability of insect populations. For the present, corn farmers seem to be winning, but some farmers think that in the long run insects may carry the day.

The McLean County farmer's year is still dictated by nature's rhythms. Much of the work for corn goes on in the fall. "We usually work after we take the beans off in the fall; we disk that ground, we apply all our fertilizers for the following year of corn" (McFIP, Ray Wesley Rafferty). Many farmers do not work the ground at all in the spring. Winter is the season for education, planning, seed selection and equipment maintenance. Seed selection is critical. In McLean County, common practice is to plant several varieties of hybrid corn seed. "It kind of spreads the risk out. Some varieties, given the year, a disease will hit that corn or whatever, maybe the bugs are worse in a particular variety and you never know what corn is going to do that. We usually plant, I would say, five different varieties of corn . . ." (McFIP, Ray Wesley Rafferty). Hybrid seed is expensive. "Seed corn right now, we bought some corn and it was right at $100 a bag. It planted three acres" (McFIP Arthur Nafziger). About 1960, McLean County farmers began to change from double cross to single hybrids. The new hybrids were better producers, but, like many thoroughbreds, more temperamental and quicker to respond to slight environmental stress. Each hybrid is also likely to require its own special mix of herbicides and fertilizers. Seed corn is also becoming increasingly specialized and the choice is never easy. For many, the final decision is still made by the time tested process of remembering which of the neighbor's fields was most admired.

Today the farmer's field work is likely to begin with an April application of herbicide. This "burn down" is designed to destroy

Figure 8:13. Cat Challenger, grain cart, and 9600 John Deere tractor harvest corn in October 1995. (Ann Charback)

all vegetation which might compete with the corn's demands for moisture and nutrients. After the burn down, farmers begin their most intense period of weather watching.

In one respect, corn farming has not changed in a hundred and fifty years. McLean County farmers agree that farming is a stressful business and that planting is the most stressful time of the year. Today most corn in McLean County is planted in the last two weeks of April, ten to fourteen days earlier than it was before hybrid corn. "Planting times, a lot of time you could miss one day and you miss maybe a month or maybe three weeks. It could be raining or whatever and that happened last spring. We got some beans planted in May and then it set in raining. Then we did not get back into the fields until about the 15th of June and we finished soybeans, I remember the 15th of June. We ran two corn planters and two bean planters and finished on a Sunday the15th day of June. It rained that night and if you didn't get it planted the next day, it was another week or so, before they got back to the field after that" (McFIP, Ray Wesley Rafferty).

Like almost all other Cornbelt machinery, corn planters have grown immensely in size and price. In 1950, a four-row planter attracted attention. By the late 1990s, 12-row planters had become the common standard, but some farmers use 16-row and 24-row machines. A 24-row planter will cost today's farmer $90,000 to $100,000. Because central Illinois farmers are not keen on buying used equipment, the area has become an important source of used planters and combines. Farmers from the fringe of the Cornbelt arrive each year in search of good equipment which is no longer quite up to heart-of-the-Cornbelt standards.

Summer brings worries about insects, some spraying, and constant weather watching. Overall, however, there is much less summer work in corn than at any time in the county's history. A gap has developed in the months which used to be devoted to cultivation. During this time, a great deal of rain is not as important as rain at the most critical time of the year. Farmers will often talk about "that million dollar rain" which comes in late July.

Harvest is the reward, and most farmer's favorite time of the year. Almost all corn is harvested by a self-propelled combine which picks, shells, and pulverizes cobs. Modern combines are reliable and impressive. Walter Weinheimer commented on modern harvesting by saying, "Well [I will] tell you, last year they came over here, started to cross the road. They added one round. There is 26 acres in that field. I went out and I crawled in the cab with him and in an hour and ten minutes we were finished up down in that corner. Started up across that field; he began to press buttons, told me how many bushels we harvested out of it, how long we had taken, took one hour and ten minutes to combine it, and what it made and

Figure 8:14. Al Bateman dumps harvested grain into a pit before drying. (Ann Charback)

what the moisture was and everything. Anything you wanted to know you just pressed a button" (McFIP). Modern combines are also extremely expensive. Carl Graf, Jr. says, "The combine cost[s] $120,000 with a platform and with the accessories. A bare down combine — I don't think you can get for $100,000 — that is list price. Now if you are trading something in and the guy is making you a deal . . . It's just like buying an automobile" (McFIP).

In the harvest field, with the combine, are wagons and trucks. "You use an auger wagon to catch the combine on the go, which increases your acreage; you can do a day almost . . . at least a third because the combine never stops and unloads on the go. We have three grain trucks. We haul this corn to our bins and of course, it goes into an auger and up into the grain bin. There it is dried and after it gets the moisture down to 14 to 15 percent moisture, we transfer it to other bins. So there is a lot of work to harvest time" (McFIP, Ray Wesley Rafferty). Harvest is much faster than it has ever been. Today five men: combine driver, auger wagon operator, and three truck drivers to shuttle corn to the elevator can expect to harvest 100 acres of corn a day.

The rewards can be considerable. In a good year, like 1996, when planting was done on time and summer rains came, many McLean County farmers averaged over 159 bushels per acre and many fields exceeded 200. Even in a terrible year, like the dry insect-infested summer of 1988, some farmers still averaged 90 bushels to the acre. As always, there is keen rivalry among farmers. One McLean County farmer remarked that he had learned "never to tell what your corn made until everybody else was through and then you told what yours made. . . . You always tried to tell last because you wanted to have at least the best" (McFIP, Merlin Eugene Tyner).

Ironically, mechanization does not mean more machinery. Today's farmer probably has fewer tools than his grandfather did. A planter, a sprayer, a combine, and a good tractor will just about do the job. And the tractor need not be one of the huge land dreadnoughts popular in the 1970s. There has been a return to modest sized tractors which are lighter and therefore do not compact the soil as much and which, of course, are less expensive.

Machinery has created new roles for farm women. Evelyn Hilton Schwoerer was born in 1925 and grew up on a farm outside Bloomington. As a child, she carried wood, fed corncobs into the cook stove, and hauled out the ashes when the cooking was done. She did chores, fed chickens, gathered eggs, helped feed the threshing crews. But she described herself as "a tomboy, tag-along, dad's girl, dad's helper." In return, her father nicknamed her "Bridgette" and patiently answered her questions. One day, when Evelyn was twelve, her father came into the house and called, "Bridgette, get ready and come out here and learn how to drive the tractor. The hired man has run through the fence twice." That was it for the hired man. Evelyn went outside, crawled up onto the seat of the John Deere, and learned how to operate it. "Of course," she recalled, "girls weren't doing that at the time. Everybody said, 'My goodness, he got your daughter out on a tractor.'" But Evelyn was ready. She had already driven horses and mules and plowed corn with a one row plow. "So now I was learning something new" (McFIP).

In 1947, Evelyn married a farmer just out of service and settled into the role of farm wife in a greatly altered system of farming. During harvest, as a young wife, she would haul grain. "I hauled in for years and unloaded the grain while he ran the harvesting equipment. . . . I was there, back and forth. It was just more comforting for me to haul it in . . . some farm women say they like to operate the combine, but I have never taken that on. [I would] just as soon drive the tractor and haul the wagons, or haul to the elevator with the truck" (McFIP).

Machinery has meant the removal of much of McLean County's agricultural past. Among the things to vanish was the farm fence. Walter Weinheimer says, "Oh yes, we had a lot of fences. That was a springtime job. That was — you were busy on rainy days fixing fences." He continues, "No don't need fences any more. Fence rows, that was another rainy day job. . . . Used to be a lot of hedge fences here. Those are all pulled out. . . . Gosh, my dad started taking some out with a steam engine. Charlie Owns from Danvers pulled out this one, south side of the road" (McFIP). There were problems associated with hedge row destruction. Hedges served as windbreaks. "Those old times was bad when they pushed all the hedges out in

Figure 8:15. Corn begins its journey to the elevator on the Beehn hog farm in the fall of 1995. (Ann Charback)

this county. Let the wind come across the prairie" (McFIP, Arthur Spencer).

Some hedges were already disappearing in the late 1930s. The 1939 report for Stevenson's Farm 46 stated, "It would be a recommendation to pull out the big hedges around the farm as fast as finances would permit. These hedges are sapping the fertility and moisture from the soil, and preventing us from growing a crop within quite some distance of each hedge fence." The 1940 report shows that $174.50 was spent on hedge pulling, which was a major expense, amounting to about a quarter of the landlord's total expense on the farm. There is a note in this report which reads "They have also pulled a big hedge back of the building, and have trimmed up all the other hedges on the farm." All over the county, farmers were doing the same thing. Between 1935 and 1970, hundreds of miles of Osage orange hedgerows vanished.

The disappearance of fences is closely related to a sharp decline in the percent of farms that raised livestock. In 1954, 87 percent of McLean County farms had cattle and about 60 percent raised hogs. As recently as 1964, over half of the farms in the county still raised cattle and over 40 percent had hogs. By 1992, only a few farms had cattle and only 10 percent raised hogs. Between 1954 and 1992, the number of cattle dropped from just over 101,254 to 15,019. During the same period, the number of hogs fell from146,430 to 84,793. What these numbers mean is that cattle and hog raising are now limited to a small number of specialized farms and they are no longer part of a typical corn farm. "Well, of course, we got away from livestock and really that's the only reason we need a fence, was for livestock. It got to the point where we really couldn't make money feeding livestock, so we quit that practice and it makes it a lot easier to farm without those fences in there. No weeds to mow or nothing, so they came out, I suppose about twenty years ago. There are very, very, few fences in McLean County today" (McFIP, Ray Wesley Rafferty).

When Merlin Tyner was growing up, there were hedge fences around the outside and through the middle of their farm. He remembers, "Well the hedgerows were taken out first, of course, because of the amount of tillable soil they used up, plus the amount of moisture they took out on both sides. If the hedge got very big you sacrificed about eight or 10 rows on the side of it due to the fact the hedge used moisture out of the ground. So the corn on eight or 10 rows on the side didn't make much" (McFIP). Tyner's other fences soon followed the hedgerows. As the livestock disappeared, the fences came out. "I don't know exactly what year that was but, well, most of the permanent fences probably came out in the middle '40s, early '50s, somewhere around there. The people just built temporary fences when we had hogs out in the field; before the confinement systems came along, why you [would] just built temporary fences. There weren't hardly any permanent ones left" (McFIP).

Figure 8:16. Darrell Beehn, Jr. hauls grain to the elevator near Shirley, Illinois, in the fall of 1995. (Ann Charback)

Another casualty of the new methods of farming was the corn crib. With the switch to combines, there was no longer any need to store ear corn. Most of the remaining cribs in McLean County now stand empty throughout the year. Merlin Tyner recalls the fate of the crib on his farm. "There was some good oak in that corn crib and we used it to make a little bit of furniture. In fact, some of the very best oak in that corn crib was used to build a tree house for our grandson" (McFIP). Arthur Nafziger remembers, "We had an old crib that my grandfather had built, held 3,000 bushels. The frame was walnut, but we couldn't salvage it because the siding had been nailed on with nails and no person . . . would ruin a planing mill on that. They wouldn't even touch it. So we just burnt it" (McFIP).

Many other buildings have become redundant. In this sense, farming in McLean County is simpler than ever before. Bigger but simpler. Dan Pleines says that, unless you do store grain on your own farm, a headache he doesn't want, a farmer could get by with only a few buildings "Oh you could get by if you wanted to with your home and a really nice machine shed" (McFIP).

The new machinery is related to the destruction of many traditional farm buildings. Visitors to rural McLean County are often impressed by the small number of structures on each farmstead. Old barns and sheds which, in other parts of the country might be adapted to accommodate machinery, are simply too small for the planting and harvesting machines now used in the central Cornbelt. Moreover, the larger the machinery, the more difficult it is to maneuver. Many of the surviving buildings serve no real economic purpose. Gregory Otto says, "There is a barn standing, but it houses mainly cats and antique machinery" (McFIP). So the old buildings go. Regrettably, because the county once boasted many large and architecturally interesting barns and cribs. Fortunately, a number of farmers are now showing interest in preserving barns as a reminder to future generations of what farming was once like.

Today, marketing corn is perhaps the farmer's toughest job. Many McLean County farmers regard it as gambling, but it is a form of gambling where farmers can do some things to alter the odds in their favor. "The farmer to me is the biggest gambler there is. We roll higher stakes than they do in Vegas" (McFIP, Evelyn Schroewer). The gambling is difficult because it is international gambling. There is keen awareness of the need for farmers to understand worldwide markets. "They have to know what's going on in South American countries, China, Japan, Russia, and the whole world. It makes a big difference in the marketing and what is going on out there. If Brazil and Argentina are having a poor bean crop, it's a sign that you want to hold onto your beans; the price will certainly go up. So they have a good crop, why it goes the other way" (McFIP, Ray Wesley Rafferty). Long ago, the price of corn reached its high point just before harvest, so many farmers tried to avoid selling at this time. But since there is so much storage now available, and because buyers and sellers are always gambling against each other on the price of corn, there are no hard and fast rules to the game of selling corn. Since the mid nineteenth century, corn futures have been available to McLean County farmers, so a crop may in effect be sold before it is even planted. But the price a farmer gets for his corn is influenced by much more than worldwide supply and demand. It also depends on broader conditions in international money markets. Today a major potential purchaser may consider stocks, bonds, or dozens of other financial instruments as alternatives to the grain market.

Where does the corn go? The simple answer is that about 60 percent of McLean County corn still goes to feed livestock. Of course, it is no longer primarily McLean County livestock. Corn winds up as chicken feed in Arkansas, hog feed in North Carolina, or it may simply be hauled to the Illinois River for shipment by barge to New Orleans and then to almost anywhere in the world. In part, it is a question of latitude. Much of Europe is too far north to be prime corn growing country. Nor are the tropics really competitive; corn is a sunshine-loving plant and does far better with 17 hours of McLean County July sunlight than with 12 hours of light astride the equator.

Many farmers would agree with Carl Graf when he says that marketing is the hardest thing about farming. "The hardest part of farming today is [the] market. It's just like this week the river. . . . Today is Monday. It rained on the river last night down in Ohio. The eastern side is going to have to deal with rain. They are going to get behind. Well, they are going to get to have an influx of a lot of corn when the farms can get going away from the river." Therefore, Carl felt that the bases, or margin between the farm and consumer, were going to fall. "You don't make a dime from corn," says Carl, "until you get to the bank and get the check cashed. You can lose it in the elevator. You can lose it in the harvesting. A farmer is the highest faith gambler in life because he plants corn with faith. And all he does, as the gambler, he increases and decreases the odds of making money" (McFIP).

Arthur Nafziger also regards marketing as the most difficult part of farming. "Pretty near everybody can raise a crop if they pay half way attention to what they do in the fields and everything — but in the last couple of years its been [a] two to three dollar swing in beans and corn. If you are at the low end, well you are at the low end and if you are at the top end you are lucky. But . . . the biggest challenge as far as I'm concerned of all the years I have farmed is knowing when to market" (McFIP).

Farmers are never happy with the way pricing decisions are made. Evelyn Schwoerer put it this way. "I would like to bring the Board of Trade members down here and show them what beans and corn look like" (McFIP). Farm families are acutely aware that the absolute price of corn is much more than buying power. Merlin Tyner remarks, "I started farming in 1946. And I sold the first crop of corn in 1946. I raised and sold it for $2.50 a bushel. I would like to be able to sell corn today that would buy what two dollars and a half did then" (McFIP). Gregory Otto, who farms near Danvers,

Figure 8:17. Delivering grain to the elevator at Shirley, Illinois, 1995. (Ann Charback)

says, "I know grandfather used to [use] a one row horse power [planter] and now we are using 16-24 row equipment. The ironic thing is I don't think the price of corn has changed much. The yield has probably doubled, but the price hasn't changed much. So how can you justify all this technology? I don't know" (McFIP).

The only way to meet the increased cost of farming was to farm more land. Some farmers regard 1,000 to 1,500 acres as the minimum for successful McLean County corn farming. Retired farmer Marlin Eugene Tyner reports that a tenant now farms his 400 acres plus another 600 acres, "He farms over a thousand acres. Of course that is one big difference; I mean, when I started farming, I suppose . . . well I think 240 acres might have been about the average, maybe not even that much when I started. More like 180. Today the average is somewhere up there around a thousand" (McFIP). Arthur Nafziger also thinks that land will be a major problem for future farmers. "You have to have so many more acres. My grandfather made a good living on 160 acres. We made a good living on 520. My son can't get enough on 520 and he has to supplement it with another job. You need pretty near to have a thousand acres today to make a living" (McFIP).

Figure 8:18. Much McLean County corn still goes for animal feed but ethanol producers have become major consumers. This sign is at the Leo Miller farm. (W. Walters)

Obtaining land has become a serious problem. Another farmer remembered, "Of course when I grew up, why, you could find a piece here and there. Of course you didn't go very far, always stayed close. People moved — first of March you could see wagons going up and down the road, somebody moving here and there every year. You don't see that any more" (McFIP, Walter Weinheimer). Moreover, the farmer is now often being outbid by non-farm users. Urban growth on the fringes of Bloomington and Normal has consumed thousands of acres of prime agricultural land. Real estate developers are also now developing dozens of residential clusters miles from the edge of the twin cities; most of these are on more wooded and more rolling land. Interstate highways consumed large amounts of prime corn ground. Most farmers accept these changes philosophically. They are angrier about another consumer of agricultural land, the growing number of gravel pits with their permanent clouds of dust and swarms of rock-spewing dump trucks. Perhaps this contrast in attitudes comes from a sense that an urban family living on the edge of a cornfield is, at least, in a limited sense, a participant in the good life of the countryside, while the man who digs for gravel is the ultimate and total destroyer of something which ought to be held in trust. Those who accept the idea that land is held in trust for future generations often ask this simple question: how do you explain to God that you have been given custody of over 700,000 acres of the world's best farmland and have somehow managed, in less than a generation, to fill substantial parts of it with holes or cover it with concrete? Between 1978 and 1992, about four percent of McLean County's farmland disappeared, and the rate of loss, has certainly been at least as great since 1992.

Those who hold this trust are fewer than ever. In 1880, 5,466 farms raised 11,976, 581 bushels of corn. Today fewer than 1,600 farms raise more than five times that amount of corn. By every measure — dollars generated, corn per acre, total production — each farmer is many times more productive. Yet, success cannot simply be measured by statistics. There are units of measure which do not fit conveniently into computers. The overall equation of McLean County corn farming is something more than tallies of bushels per acre or sums on a balance sheet. It would be unwise to finish the account of corn raising in the heart of the Cornbelt without listening to farmers when they speak about the intangibles of farm life.

Many farmers simply enjoy the act of farming. Often they have spent a good deal of time thinking about it, and it is not uncommon to find farmers who will admit to being part-time philosophers. When they are asked what they enjoy about farming, they usually have an answer. Most often, this answer comes in the repeated use of a single word: freedom. Evelyn Schwoerer replied.

Figure 8:19. Murphy Farms near Shirley, Illinois. (Ann Charback)

"Freedom. You do what has to be done that day and if you run over 15 minutes on the job you are not going to get condemned. You do a job and you do it right. No time limits. More relaxed. Have to get it done, you know you have to do it, but you set your own time frame" (McFIP). Ray Wesley Rafferty put it this way, "Well, I think that [it] is a great way of life. You're away from the hustle, bustle of city traffic, lots of people. You're out there pretty much on your own as far as you can do what you want, when you want to, if you want to and the more you do, the better job you do, the more you are going to make and the better living you are going to have for it" (McFIP). Others reflect this theme of responsibility. "The greatest thing today about farming probably is the fact that you are more nearly your own boss than anywhere else you can work. You make your own decisions and you pay for your own mistakes, even though everybody likes to think that whatever happens is somebody else's fault. If you're a farmer, more than likely, it isn't" (McFIP, Merlin Eugene Tyner).

Freedom is linked to pride. Arthur Spencer says, "Well one thing I think most farmers like about being a farmer is their independence. They like that freedom. They make their choices. Like picking — starting out — like picking their hybrids, picking their beans. They get to do that themselves. They get to sell their product. Sometimes they get a little discouraged, unhappy with the market, but the independence is there and the freedom. Another thing, when they are out there working, there is nobody bossing them. They are doing their own thing. They know what they want to do. They study the market. They study their seed corn. They have studied how to no-till and all that stuff. They take a world of pride in [it]" (McFIP).

Along with freedom, family and nature and friendship are important. Arthur Nafziger says the best thing about farming is, "Being with nature, seeing the pheasants, seeing foxes, birds and the wrath of God when storms come up and the goodness of God when he sends you rain. It is just a great way to raise a family. It is up to you how much you want to work. If you want to go to town Tuesday afternoon . . . you go to town on a Tuesday afternoon. You do have that freedom." He continued, "It's much freer, the neighbors I found in town, it was a little harder to get to know your neighbors than it is in the country. If I was out in the field, even with a tractor or everything, I'd come up to the fence and the neighbor on the other side came up to the fence about the same time, we always stopped and talked a little bit. Things like that; and if we weren't busy we talked a lot. Then also, any neighbor that had an accident or anything, everybody would plow for people, they would combine, they would pick corn or put in crops and things like that. Any neighbor that had an accident — I know one fellow he got leukemia — we had the elevator closed to everybody but that fellow and we had about 20 combines and 30 trucks that day. We combined I don't know how many acres for this fellow. Took it to the elevator, like I say the elevator was closed to anybody, and nobody else could bring corn in but this one place so we could get him done in one day" (McFIP).

Scott Hoeft agrees. "Yes. It is a real close knit community. My dad died in May and we had the corn planted but we didn't have the beans planted. All the neighbors came in and plant[ed] the beans for us. We were building this building out there. We didn't have any of the tin on and they all got together one day and just come. They put up the building. I left one morning and come back and it was done" (McFIP).

When asked what skills were needed to become a successful farmer, Mark Deterding replied, "Well, patience, I would say, mechanical ability. Optimism. I guess those are the biggest things. Figuring that you are not going to get rich. Just enjoy the out of doors is probably the most important thing. 'Cause that is where you are going to be doing the work." He goes on to reflect on his boyhood growing up on a farm. "It was good. There was always something to do. We worked every day and I don't mean work that you wouldn't like to do. We did a lot of work but we also had a lot of fun doing it. I wouldn't trade it for anything" (McFIP).

REFERENCES AND BRIEF BIBLIOGRAPHY

UNPUBLISHED MANUSCRIPTS

(McLean County Historical Society unless otherwise noted.)

Baldridge, Samuel.
Memoirs.

Barnard, Osborne.
Diary 1860.

Benjamin, William H.
Diary 1858.

Bowman, Archie.
Undated typescript.

Britt, Richard M.
Diary 1865.

Corn Belt Exposition 1938.
Program.

Corn Bowl
Programs 1947, 1948.

Federal Surveyors' Field Notes and Plats.
McLean County Recorder of Deeds.

Heafer, Edgar.
"Drain Tile – Historical Sketch."

Hendricks, Melvin.
"Days I remember."

Koos, Greg.
1995. "Farmer Origin and Farm Production in McLean County, Illinois, 1850".

McClun, George Edward.
Papers, Illinois State University Archives.

McFIP.
McLean County Farm Interview Project, 1997.

"Prairie Farming"
1855

Schilt, Arlene Rose.
1972. Noble-Weiting: An Early Upper Mississippian Village. Masters Thesis, Illinois State University.

Shirley Corn and Horse Show.
Program.

Smith, Charles Raymond.
1978. The Grand Village of the Kickapoo. Masters Thesis, Illinois State University.

Adlai Stevenson Family farm reports.
Various dates.

Warren, Horace.
1856. Letter.

Walters, William D., Jr.
1973. "Historic Buildings of McLean County."

BOOKS and PAMPHLETS

Arthur, Eric and Dudley Whitney.
1972. *The Barn: A Vanishing Landmark in North America*. Toronto: A & W Visual Library.

Bogue, Margaret B.
1954. *Patterns From the Sod: Land Use and Land Tenure in the Grand Prairie 1850-1900*. Springfield: Illinois State Historical Society Library.

Bowman, M. L. and B. W. Crosley.
1908. *Corn*. Ames, Iowa: Bowman and Crosley.

Burnham, J. and Ezra Prince.
1908. *Historical Encyclopedia of Illinois and History of McLean County*. Chicago: Munsell.

Cavanaugh, Helen.
1959. *Seed, Soil and Science: The Story of Eugene D. Funk*. Chicago R. R. Donneley.

Cavanaugh, Helen.
1952. *Funk of Funk's Grove*. Pantagraph Printing Co.

Conzen, Michael P.,ed.
1990. *The Making of the American Landscape*. Boston: Hyman.

Crabb, Richard A.
1993. *The Hybrid Corn Makers*. 2nd ed. Wheaton, Illinois. Richard Crabb.

Cronon, William.
1991. *Nature's Metropolis: Chicago and the Great West*. New York and London: W. W. Norton.

Drury, John.
1942. *Old Illinois Houses*. Springfield: Occasional Publications of the Illinois Historical Society.

Duis, E.
1874. *The Good Old Times in McLean County, Illinois*. Bloomington: Leader Publishing and Printing.

Faragher, John Mack.
1986. *Sugar Creek: Life on the Illinois Prairie*. New Haven: Yale University Press.

Fitzgerald, Deborah.
1900. *The Business of Breeding: Hybrid Corn in Illinois, 1890-1940*. Ithaca & London: Cornell University Press.

Hardeman, Nicholas P.
1981. *Shucks, Shocks, and Hominy Blocks: Corn as a Way of Life in Pioneer America*. Baton Rouge and London: Louisiana State University Press.

Hart, John Frasier.
1991. *The Land That Feeds Us*. New York: W. W. Norton.

Hartley, C. T.
1915. *Corn Cultivation*. Farmer's Bulletin 414. Washington, DC: Government Printing Office.

Hasbrouck, Jacob L.
1924. *History of McLean County Illinois*. 2 Vols; Topeka & Indianapolis: Historical Publishing Company.

Hayes, Herbert Kendall.
1963. *A Professor's Story of Hybrid Corn*. Minneapolis: Burgess.

History of McLean County Illinois.
1879. Chicago: William Le Baron.

Hudson, John C.
1994. *Making the Cornbelt: A Geographical History of Middle Western Agriculture*. Bloomington: Indiana University Press.

Hurt, Douglas.
1982. *Early Farm Tools: From Hand Power to Steam Power*. Yuma, AZ and Manhattan, KS: Sunflower University Press.

Jesse, James W.
1997. *Civil War Diaries of James W. Jesse 1861-1865*. William P. LaBounty, editor; Transcribed by Alan D. Selig. Bloomington. McLean County Genealogical Society.

Johnson, William C.
1844. *The Farmer's Encyclopedia*. Philadelphia: Carey & Hart.

Kaufman, Henry J.
1965. *The American Farmhouse*. New York: Bonanza.

Klampkin, Charles.
1973. *Barns: Their History Preservation and Restoration*. New York: Bonanza.

Koos, Greg; Don Munson, and Martin A. Wycoff.
1982. *The Illustrated History of McLean County*. Bloomington: McLean County Historical Society.

Lewis, Wiley B.
1970. *Corn Pickers and Picking Corn*. Columbus: Ohio Agricultural Education Curriculum Materials Service.

McManis, Douglas.
1964. *Initial Evaluation and Utilization of the Illinois Prairies, 1815-1840*. Chicago: University of Chicago Press.

Metcalf, Robert A.
1993. *Destructive and Useful Insects*. New York: McGraw Hill.

Moore, Arthur.
1945. *The Farmer and the Rest of Us*. Boston: Little Brown.

Myrick, Herbert, ed.
1903. *The Book of Corn*. New York & Chicago: Orange Judd.

Noble, Allen and Margaret M. Geib.
1984. *Wood, Brick, and Stone: The North American Settlement Landscape*. Amherst: University of Massachusetts Press.

Oglesby, Richard. 1894.
Impromptu Speech of Ex-Gov. Richard Oglesby. Written from memory by Volney Foster, 1898.

Oliver, William.
1843. *Eight Months in Illinois: With Information for Immigrants*. Newcastle Upon Tyne: Thomas Mitchell.

Partridge, Michael.
1972. *Farm Tools Through the Ages*. Reading, UK: Osprey.

Portrait and Biographical Album of McLean County, Ill.
1887. Chicago: Chapman Brothers.

Roe, Keith.
1988. *Corncribs: In History, Folklife, and Architecture*. Ames: Iowa State University Press.

Thompson, Dave. ed.
1964. *Five Golden Decades*. Bloomington: McLean County Farm Bureau.

Stevenson, A. E.
1887. Address Before the Agricultural Association of the Fifth Congressional District of Illinois. Washington, DC: Thomas McGill.

Sublett, Michael D.; William D. Walters, Jr. and Suthard Modry.
1973. *Commentary of a Cornbelt Countryside: A Self-Guided Rural Experience*. Normal: Department of Geography-Geology.

ARTICLES

Borland, Thomas.
"The Wire Fence," *Prairie Farmer* 10:4, 113-114.

Carney, George O.
1995. "Grain Elevators in the United States and Canada: Functional or Symbolic," *Material Culture* 27:1, 1-24.

Catton, J. D.
1841. "Breaking Prairie," *Union Agriculturist* 2:5 34-35.

Crandel, J.
1842. "The Culture of Corn" *Union Agriculturist* 2:5, 43.

"The Canadian Excursion"
1860. *Prairie Farmer* 20:8, 81-82.

Cannon, Brian G.
1991. "Immigrants in American Agriculture" *Agricultural History* 65:1, 17-35.

"Corn Culture."
1868. *Prairie Farmer* 21:11, 14.

Dement, Issac T.
1853. "Hedge Culture" *Prairie Farmer* 13:2, 81.

"Drainage."
1852. *Prairie Farmer* 12:2, 26.

"Draining."
1841. *Union Agriculturist* 1:7, 50.

Ekberg, Carl.
1995. "Agriculture, Mentalities, and Violence on the Illinois Frontier," *Illinois Historical Journal* 88:2, 101-116.

Funk, Henry.
1868. "How Shall We Manure?" *Prairie Farmer* 21:14, 218.

Gersmehl, Philip J.
1978. "No-Till Farming: The Regional Applicability of a Revolutionary Agricultural Technology" *Geographical Review* 68:1, 66-79.

Hall, N. E.
1847. "Fence Experiments" *Prairie Farmer* 7:3, 98.

Hall, N. E.
1847. "Prairie Breaking" *Prairie Farmer* 7:9, 317.

Hart, John Fraser.
1986. "Change in the Corn Belt," *Geographical Review* 76:1, 51-72.

Hamilton, Henry W. and Jean Tyree Hamilton.
1978. "Alfred Montgomery: Itinerant Midwestern Artist," *Bulletin* Missouri Historical Society XXIV:2 69-82.

"How We Have All Advanced."
1868. *Prairie Farmer* 21:2, 17.

Kidder, Henry M.
1865. "Land Drainage," *Transactions of the Illinois State Agricultural Society* V 547-552.

"Land Draining."
1857. *The Valley Farmer* 9:2, 47-49.

Rezelman, John.
1983. "Consider the Corn Shock" *Country Journal* X:9, 60-67.

Walters, William D., Jr.
1979-1980. "Abandoned Nineteenth Century Brick and Tile Works in Central Illinois: An Introduction from Local Sources" *Industrial Archaeology Review* 4:1, 70-80

Walters, William D., Jr. and Floyd Mansburger.
1982. "Initial Field Location in Illinois," *Agricultural History* 57 289-296.

Walters, William D., Jr. and Jonathan Smith.
1992. "Woodland and Prairie Settlement in Illinois: 1830-1870," *Forest and Conservation History* 36:1, 17-21.

Winsor, Roger A.
1987. "Environmental Imagery of the Wet Prairie of Central Illinois, 1820-1920," *Journal of Historical Geography* 13:4, 375-395.

INDEX